Methods in Molecular Biology

Series Editor
John M. Walker
School of Life and Medical Sciences
University of Hertfordshire
Hatfield, Hertfordshire, UK

For further volumes:
http://www.springer.com/series/7651

For over 35 years, biological scientists have come to rely on the research protocols and methodologies in the critically acclaimed *Methods in Molecular Biology* series. The series was the first to introduce the step-by-step protocols approach that has become the standard in all biomedical protocol publishing. Each protocol is provided in readily-reproducible step-by-step fashion, opening with an introductory overview, a list of the materials and reagents needed to complete the experiment, and followed by a detailed procedure that is supported with a helpful notes section offering tips and tricks of the trade as well as troubleshooting advice. These hallmark features were introduced by series editor Dr. John Walker and constitute the key ingredient in each and every volume of the *Methods in Molecular Biology* series. Tested and trusted, comprehensive and reliable, all protocols from the series are indexed in PubMed.

Splice Switching

Methods and Protocols

Edited by

Yoshitsugu Aoki

*Department of Molecular Therapy, National Institute of Neuroscience,
National Center of Neurology and Psychiatry, Tokyo, Japan*

Norio Motohashi

Center of Neurology and Psychiatry, National Institute of Neuroscience, Tokyo, Japan

Humana Press

Editors
Yoshitsugu Aoki
Department of Molecular Therapy
National Institute of Neuroscience,
National Center of Neurology and Psychiatry
Tokyo, Japan

Norio Motohashi
Center of Neurology and Psychiatry
National Institute of Neuroscience
Tokyo, Japan

ISSN 1064-3745 ISSN 1940-6029 (electronic)
Methods in Molecular Biology
ISBN 978-1-0716-4737-0 ISBN 978-1-0716-4738-7 (eBook)
https://doi.org/10.1007/978-1-0716-4738-7

This Humana imprint is published by the registered company Springer Science+Business Media, LLC, part of Springer Nature.
The registered company address is: 1 New York Plaza, New York, NY 10004, U.S.A.

Preface

The *Methods and Protocols* outlined in this volume reflect the rapid advancements in the field of RNA therapeutics, with a particular focus on the development of splice switching, exon skipping, and inclusion therapies as targeted treatments for genetic diseases, as well as the evaluation methods used post-treatment. Remarkable progress has been made in understanding alternative splicing mechanisms and designing therapeutic approaches that modify splicing patterns. These approaches have proven promising for several diseases, including Duchenne muscular dystrophy (DMD), spinal muscular atrophy (SMA), and other genetic disorders characterized by abnormal splicing events.

This volume discusses the latest methodologies, protocols, and new perspectives from experts who have contributed to significant breakthroughs in the field. By including both established methods and novel techniques, we aim to offer a comprehensive resource for researchers working to develop therapies that apply the potential of exon skipping and inclusion therapies.

The protocols presented in this book are designed to enable researchers across various fields to apply these techniques in their studies. The purpose is to advance basic RNA biology research or accelerate therapeutic development, and we believe this volume will provide valuable insights.

We are deeply grateful to the contributors for sharing their expertise and for their ongoing efforts to advance this exciting field. We hope this book will inspire further innovation and collaboration in exploring more effective RNA-based therapies.

Tokyo, Japan

Yoshitsugu Aoki
Norio Motohashi

Contents

Contributors

MATTHEW S. ALEXANDER • *Department of Pediatrics, Division of Neurology at Children's of Alabama and the University of Alabama at Birmingham, Birmingham, AL, USA; Department of Genetics, University of Alabama at Birmingham, Birmingham, AL, USA; UAB Center for Exercise Medicine (UCEM), Birmingham, AL, USA; UAB Civitan International Research Center (CIRC), Birmingham, AL, USA; UAB Center for Neurodegeneration and Experimental Therapeutics (CNET), Birmingham, AL, USA*

YOSHITSUGU AOKI • *Department of Molecular Therapy, National Institute of Neuroscience, National Center of Neurology and Psychiatry, Tokyo, Japan*

YUKI ASHIDA • *Department of Molecular Therapy, National Institute of Neuroscience, National Center of Neurology and Psychiatry, Tokyo, Japan*

TOMONARI AWAYA • *Center for Anatomical Studies, Graduate School of Medicine, Kyoto University, Kyoto, Japan*

YUSUKE ECHIGOYA • *Laboratory of Biomedical Science, Department of Veterinary Medicine, College of Bioresource Sciences, Nihon University, Fujisawa, Kanagawa, Japan; Nihon University Veterinary Research Center, Fujisawa, Kanagawa, Japan*

KATHERINE G. ENGLISH • *Department of Pediatrics, Division of Neurology at Children's of Alabama and the University of Alabama at Birmingham, Birmingham, AL, USA*

MASATOSHI HAGIWARA • *Department of Drug Discovery Medicine, Graduate School of Medicine, Kyoto University, Kyoto, Japan*

UMME SABRINA HAQUE • *Department of Medical Genetics, Faculty of Medicine and Dentistry, University of Alberta, Edmonton, AB, Canada*

YUKINE ISHIBASHI • *Department of Materials Engineering, Graduate School of Engineering, The University of Tokyo, Bunkyo-ku, Tokyo, Japan*

MAYUMI KAMADA • *School of Frontier Engineering, Kitasato University, Kanagawa, Japan*

YOSUKE KAWAI • *National Institute of Global Health and Medicine, Japan Institute for Health Security, Tokyo, Japan*

TAKAHIRO KAWASHIMA • *Department of Information Medicine, National Institute of Neuroscience, National Center of Neurology and Psychiatry, Kodaira, Tokyo, Japan*

KOICHI KIMURA • *Department of Cardiology, The University of Tokyo, Tokyo, Japan; Department of Molecular Therapy, National Center of Neurology and Psychiatry, Tokyo, Japan*

YOSHIHIRO KITA • *Life Sciences Core Facility, Graduate School of Medicine, The University of Tokyo, Bunkyo-ku, Tokyo, Japan*

MELISSA KOHUT • *Department of Medical Genetics, Faculty of Medicine and Dentistry, University of Alberta, Edmonton, AB, Canada*

HIROSHI KUNIISHI • *Division of Development of Mental Functions, Research Center for Child Mental Development, University of Fukui, Fukui, Japan*

KATSUHIKO KUNITAKE • *Department of Molecular Therapy, National Institute of Neuroscience, National Center of Neurology and Psychiatry (NCNP), Kodaira, Tokyo, Japan*

MUTSUKI KURAOKA • *Laboratory of Experimental Animal Science, Nippon Veterinary and Life Science University, Musashino, Tokyo, Japan; Department of Molecular Therapy,*

National Institute of Neuroscience, National Center of Neurology and Psychiatry, Kodaira, Tokyo, Japan

CHUN YIN JERRY LAU • *Department of Materials Engineering, Graduate School of Engineering, The University of Tokyo, Bunkyo-ku, Tokyo, Japan*

KATSURA MINEGISHI • *Department of Molecular Therapy, National Institute of Neuroscience, National Center of Neurology and Psychiatry (NCNP), Kodaira, Tokyo, Japan*

KANJIRO MIYATA • *Department of Materials Engineering, Graduate School of Engineering, The University of Tokyo, Bunkyo-ku, Tokyo, Japan*

HIROYUKI MORITA • *Department of Cardiology, The University of Tokyo, Tokyo, Japan*

NORIO MOTOHASHI • *Center of Neurology and Psychiatry, National Institute of Neuroscience, Tokyo, Japan*

MITSURU NAITO • *Department of Materials Engineering, Graduate School of Engineering, The University of Tokyo, Bunkyo-ku, Tokyo, Japan; Faculty of Advanced Engineering, Tokyo University of Science, Tokyo, Japan*

MASAYUKI NAKAMORI • *Department of Neurology, Yamaguchi University Graduate School of Medicine, Yamaguchi, Japan*

YUKO NITAHARA-KASAHARA • *Division of Molecular and Medical Genetics, Center for Gene and Cell Therapy, The Institute of Medical Science, The University of Tokyo, Minato-Ku, Tokyo, Japan*

HIROAKI OHARA • *Department of Drug Discovery Medicine, Graduate School of Medicine, Kyoto University, Kyoto, Japan*

HIROYA ONO • *Department of Molecular Therapy, National Institute of Neuroscience, National Center of Neurology and Psychiatry (NCNP), Tokyo, Japan*

JOHN M. PARANT • *Department of Pharmacology and Toxicology, University of Alabama at Birmingham School of Medicine, Birmingham, AL, USA*

HIDETOSHI SAKURAI • *Center for iPS Cell Research and Application (CiRA), Kyoto University, Kyoto, Japan*

MD NUR AHAD SHAH • *Department of Medical Genetics, Faculty of Medicine and Dentistry, University of Alberta, Edmonton, AB, Canada*

HIDEO SHINDOU • *Department of Medical Lipid Science, Graduate School of Medicine, The University of Tokyo, Bunkyo-ku, Tokyo, Japan; Department of Lipid Life Science, Japan Institute for Health Security, Shinjuku-ku, Tokyo, Japan*

LAURA EDELLEIN SUTANTO • *Department of Medical Genetics, Faculty of Medicine and Dentistry, University of Alberta, Edmonton, AB, Canada*

HISATERU TACHIMORI • *Department of Information Medicine, National Institute of Neuroscience, National Center of Neurology and Psychiatry, Kodaira, Tokyo, Japan; Department of Health Policy and Management, Keio University School of Medicine, Shinjuku-ku, Tokyo, Japan*

SHIN'ICHI TAKEDA • *National Institute of Neuroscience, National Center of Neurology and Psychiatry, Kodaira, Tokyo, Japan*

ERI TAKEUCHI • *Department of Molecular Therapy, National Institute of Neuroscience, National Center of Neurology and Psychiatry (NCNP), Tokyo, Japan*

HOLLY R. THOMAS • *Department of Pharmacology and Toxicology, University of Alabama at Birmingham School of Medicine, Birmingham, AL, USA*

SUZUMI M. TOKUOKA • *Department of Lipidomics, Graduate School of Medicine, The University of Tokyo, Bunkyo-ku, Tokyo, Japan*

TSUKASA TOMINARI • *Department of Molecular Therapy, National Institute of Neuroscience, National Center of Neurology and Psychiatry, Kodaira, Tokyo, Japan*

TOMOYA UCHIMURA • *Center for iPS Cell Research and Application (CiRA), Kyoto University, Kyoto, Japan*
YUKIKO UETA • *Department of Drug Discovery Medicine, Graduate School of Medicine, Kyoto University, Kyoto, Japan*
WILLIAM J. VALENTINE • *Department of Molecular Therapy, National Institute of Neuroscience, National Center for Neurology and Psychiatry (NCNP), Kodaira, Tokyo, Japan; Institute of Microbial Chemistry (BIKAKEN), Shinagawa, Tokyo, Japan; Department of Lipid Life Science, Japan Institute for Health Security, Shinjuku-ku, Tokyo, Japan*
KARIN WATANABE • *Department of Molecular Therapy, National Institute of Neuroscience, National Center of Neurology and Psychiatry, Kodaira, Tokyo, Japan; Graduate School of Applied Biochemistry, Nippon Veterinary and Life Science University, Musashino, Tokyo, Japan*
TAKASHI YAMADA • *Graduate School of Biomedical and Health Sciences, Hiroshima University, Hiroshima, Japan*
TOSHIFUMI YOKOTA • *Department of Medical Genetics, Faculty of Medicine and Dentistry, University of Alberta, Edmonton, AB, Canada; The Friends of Garrett Cumming Research & Muscular Dystrophy Canada HM Toupin, Neurological Science Research Chair, Edmonton, AB, Canada*

Chapter 1

DG9-Conjugated Morpholino-Based Exon 51-Skipping Therapy for Duchenne Muscular Dystrophy

Md Nur Ahad Shah, Laura Edellein Sutanto, and Toshifumi Yokota

Abstract

Duchenne muscular dystrophy (DMD) is a severe genetic disorder caused by mutations in the DMD gene that leads to the loss of the dystrophin protein. Exon-skipping therapy with phosphorodiamidate morpholino oligomers (PMOs) has been shown to restore the disrupted open-reading frame to produce a shortened yet functional dystrophin. The applications of this treatment, however, are limited in terms of its efficiency. A potential solution to this problem is to conjugate the PMOs to cell-penetrating peptides. DG9 is a novel peptide that has exhibited promising results in terms of its ability to enhance the cellular uptake of PMOs to increase exon-skipping efficiency in both the skeletal and heart muscles. In this chapter, we describe the systemic injection of DG9-conjugated PMOs to induce the skipping of exon 51 in an exon 52-deleted mouse model, h*DMD*del52;*mdx*, which is a humanized dystrophic mouse model that mimics human DMD patients by having a human *DMD* gene integrated into them with a disrupted reading frame. As a result, this genetic makeup makes the model useful for testing the safety and efficacy of DG9-PMO as a potential treatment for DMD. We also describe the methodology to assess the efficacy and safety of the DG9-PMO treatment, including systematic DG9-PMO injections, RT-PCR, western blotting, histology, and functional tests.

Key words Duchenne muscular dystrophy (DMD), Phosphorodiamidate morpholino oligomers (PMOs), Antisense oligonucleotides, DG9-PMO, Peptide-conjugated PMO, Dystrophin, Exon skipping, Dystrophic cardiomyopathy, Humanized dystrophic mouse model

1 Introduction

Duchenne muscular dystrophy (DMD) is considered to be the most common muscular dystrophy in the world, affecting about 1 in 5000 newborn males worldwide [1]. DMD is a severe X-linked recessive disorder caused by out-of-frame mutations in the DMD gene that result in the loss of dystrophin protein. Dystrophin is an essential component of the muscle as it structurally links the cytoskeletal actin to the extracellular matrix through the dystrophin-associated protein complex (DAPC), essentially stabilizing the plasma membrane of the muscle cell [1]. Since the DAPC plays a

Yoshitsugu Aoki and Norio Motohashi (eds.), *Splice Switching: Methods and Protocols*, Methods in Molecular Biology, vol. 2963, https://doi.org/10.1007/978-1-0716-4738-7_1,

structural role in maintaining the integrity of the muscle cell, the breakdown of DAPC induces body-wide muscle degeneration and cardiorespiratory complications [1]. In DMD patients, the sarcolemma is weakened due to the disassembly of DAPC, which renders it vulnerable to damage caused by muscle contractions. In DMD, about 60–70% of mutations are deletions, 20% have nonsense mutations, small insertions, or deletion mutations, and 5–15% of patients have duplications [1]. The deletion of exon 52 is the most prevalent form of deletion mutation, which encompasses about 13% of DMD patients [2]. Though there are some symptomatic treatments available, targeted toward managing the complications presented by DMD, a cure is still out of reach.

Exon skipping is a therapeutic strategy that directly targets the root cause of DMD by restoring the dystrophin protein. This approach utilizes antisense oligonucleotides, specifically phosphorodiamidate morpholino oligomers (PMOs), which are small DNA-like molecules that have been developed to specifically target an exon during pre-mRNA processing, inherently blocking the incorporation of the exon of interest into the mature mRNA [3]. This skipping of certain exons, in turn, results in the restoration of the reading frame to produce a shortened yet functional dystrophin [4]. In the case of exon 52 deletion, exon 51-skipping therapy with a PMO has shown to be able to restore the expression of dystrophin in skeletal muscles in dystrophic mouse models, along with the recovery of body-wide muscle functions [2, 5]. Here, the PMO-mediated exon 51 skipping restored the distorted open-reading frame caused by exon 52 deletion to give rise to a phenotype comparable to Becker muscular dystrophy (BMD)—a milder form of muscular dystrophy resulting from in-frame deletions in the *DMD* gene [5, 6]. Currently, the US Food and Drug Administration has approved four exon-skipping therapies that utilize PMOs [5]. Sarepta Therapeutics developed Eteplirsen/Exondys 51, Golodirsen/Vyondys 53, and Casimersen/Amondys 45, which specifically target exon 51, 53, and 45 of the *DMD* gene, respectively [7–9]. Viltolarsen/Viltepso is designed by NS Pharma and binds to exon 53 of the *DMD* transcript, inducing the skipping of exon 53 during pre-mRNA splicing [10]. These therapeutic approaches can be systematically administered in humans due to the small nature of antisense oligonucleotides.

PMOs enter the cell by diffusing into the leaky myocytes or through receptor-mediated endocytosis to modify the DMD reading frame and produce a relatively functional dystrophin, which further improves skeletal muscle functions [11, 12]. Other than targeting dystrophin restoration, symptomatic treatments like corticosteroids and angiotensin-converting enzyme (ACE) inhibitors are also prescribed to DMD patients. These drugs prevent the inflammation of the muscle tissues and reduce the cardiac workload to delay the development of dystrophic cardiomyopathy, respectively [13]. The applications of the currently approved exon-skipping therapies, however, are limited, as they have very limited efficacy in the skeletal muscle and no efficacy in the heart [1, 14]. Due to the nature of PMOs, they are almost immediately cleared from the bloodstream and have decreased uptake in the heart muscles, where most of them get trapped within the endosomes [5, 15]. Conjugating PMOs to cell-penetrating peptides has been shown to enhance the cellular uptake of PMOs in vivo to improve the efficiency of exon skipping, especially within the heart muscles [5]. DG9 is a novel peptide that has recently been shown to induce more efficient exon skipping within the skeletal and cardiac muscles upon its conjugation to PMOs [5, 19, 20]. Further testing of DG9-PMO was conducted in humanized dystrophic mouse model (h*DMD*del52;*mdx*)—a mouse model that has a human *DMD* transgene integrated into chromosome 5 with an out-of-frame deletion of exon 52 [16, 17]. The mouse model also carries the *mdx* mutation, which is a specific point mutation in exon 23 of the murine *Dmd* gene. Post-treatment analyses revealed an improved efficacy of exon 51 skipping compared to nonconjugated PMOs with a significantly increased restoration of dystrophin [5]. The widespread production of dystrophin-positive fibers was confirmed to be present in the tibialis anterior, diaphragm, and heart muscles of the DG9-PMO-treated mice through immunofluorescence analysis.

This chapter describes the method of systemic injection of DG9-conjugated PMOs that can induce exon 51 skipping in a humanized dystrophic mouse model. Systemic injections of DG9-conjugated PMO were administered to h*DMD*del52;*mdx* mice. As the onset of muscle necrosis in dystrophic mice becomes apparent around the age of 3 weeks [18], the DG9-PMO treatment was initiated at 2 months of age such that the curative potential of the treatment can be determined. Molecular and functional analysis revealed DG9-PMO treatment to be effective at restoring body-wide dystrophin protein production and improving muscle functions. As such, DG9-conjugated PMO holds a promising future that can significantly increase the efficacy of PMO-mediated exon skipping in both skeletal and cardiac muscles.

2 Materials

2.1 Animal Models

1. *Mdx* mice with exon 52-deleted human *DMD* gene integrated into chromosome 5.
2. Wild-type mice.

2.2 Administering the Morpholino Drugs

1. Antisense morpholino drug. We used the Ex51_Ac0 PMO with the sequence 5′-GTGTCACCAGAGTAACAGTCTGAGTAGGAG-3′ to skip h*DMD* exon 51.
2. Phosphate-buffered saline (PBS).
3. 1 mL syringe.

2.3 Functional Testing

1. Stopwatch.
2. Chatillon DFE II grip strength meter (Columbus Instruments).
3. AccuRotor 4-channel rotarod (Omnitech Electronics, Inc.).
4. Exer 3/6 animal treadmill (Columbus Instruments).

2.4 Tissue Collection

1. Carbon dioxide (CO_2).
2. Veterinary surgical instruments: forceps, scalpels, and scissors.
3. Tragacanth gum.
4. Cork disks.
5. Liquid nitrogen.
6. Isopentane.
7. Metal container.
8. Dry ice.
9. 25 mL syringe.
10. Falcon tubes.

2.5 RNA Extraction and RT-PCR

1. Trizol (Thermo Fisher).
2. Chloroform.
3. Isopropanol.
4. 75% ethanol.
5. RNase-free glycogen.
6. SuperScript™ IV Reverse Transcriptase (Invitrogen).
7. Random hexamers (Invitrogen).
8. GoTaq® Green Master Mix (Promega).
9. Primers for h*DMD* exon 51 skipping:
   - (a) Forward primer in exon 49/50: 5′- CAGCCAGTGAAGAGGAAGTTAG-3′.
   - (b) Reverse primer in exon 53: 5′- CCAGCCATTGTGTTGAATCC-3′.
10. Primers for m*Gapdh*:
    - (a) Forward primer: 5′- CAACTTTGGCATTGTGGAAGG-3′.
    - (b) Reverse primer: 5′- GAAGAGTGGGAGTTGCTGTT-3′.

11. RNase-free water.
12. 1.5% agarose gel.
13. GelRed nucleic acid gel stain (Biotium).

2.6 Protein Extraction and Western Blot

1. Lysis buffer: 10% SDS, 70 mM Tris-HCl at pH 6.7, 5 mM EDTA at pH 8.0, 5% β-mercaptoethanol, and cOmplete protease inhibitor cocktail (Roche).
2. Pierce™ Coomassie (Bradford) kit (Thermo Fisher).
3. NuPAGE™ LDS sample buffer 4× (Invitrogen).
4. NuPAGE™ sample reducing agent 10× (Invitrogen).
5. NuPAGE™ 3–8% Tris-acetate Midi gels (Invitrogen).
6. NuPAGE™ Tris-acetate SDS running buffer (20×) (Thermo Fisher).
7. HiMark™ prestained protein standard (Thermo Fisher).
8. Polyvinylidene difluoride (PVDF) membrane (Millipore).
9. Buffers:
   (a) Concentrated anode buffer: 0.3 M Tris–HCl, 20% methanol.
   (b) Anode buffer: 0.03 M Tris–HCl, 20% methanol.
   (c) Cathode buffer: 25 mM Tris–HCl, 20% methanol, 40 mM 6-amino-n-hexanoic acid, 0.01% SDS.
10. 2% ECL prime blocking agent (GE Healthcare).
11. PageBlue protein staining solution (Thermo Scientific).
12. NCL-DYS1 anti-dystrophin antibody (Leica).
13. Anti-desmin antibody (Abcam, ab8592).
14. PBS containing 0.05% Tween 20 (PBST).
15. Horseradish peroxidase (HRP)-conjugated anti-mouse IgG2a antibody (Invitrogen).
16. HRP-conjugated anti-rabbit IgG (H + L) secondary antibody (Invitrogen).
17. ECL select detection reagent (GE Healthcare).
18. Image Lab™ software, v.6.0.1 (Bio-Rad).

2.7 Dystrophin Immunofluorescence

1. Poly-L-lysine-coated slides.
2. Coverslips.
3. PBS.
4. Goat serum.
5. Triton X-100 (Sigma-Aldrich).
6. NCL-DYS1 anti-dystrophin antibody (Leica).

7. Alexa 488-conjugated goat anti-mouse IgG2a secondary antibody (Life Technologies).
8. Vectashield HardSet Antifade Mounting Medium with DAPI (Vector Laboratories).

2.8 Histology

1. Poly-L-lysine-coated slides.
2. Coverslips.
3. Mayer's haematoxylin (Electron Microscopy Sciences).
4. Eosin Y (Electron Microscopy Sciences).
5. Ethanol.
6. Xylene substitute (Thermo Fisher).
7. Permount™ (Fisher Scientific).
8. Nail polish.

3 Methods

3.1 Administration of Morpholino Drugs

1. Anesthetize the mice according to the animal usage guidelines.
2. Keep the mouse under anesthesia using a nose cone and administer retro-orbital injections using an insulin syringe. For repeated treatment, inject the appropriate amount of morpholino drug or PBS once per week (Fig. 1a). *See* **Notes 1** and **2**.
3. After injection, remove the nose cone and wait until the mouse regains consciousness before putting them back into the cage. Observe for abnormal behavior before proceeding to the next mouse.
4. Sacrifice and dissect the mice for tissue collection following muscle function tests, as described in the *Functional Testing* section.

3.2 Functional Testing

1. Record the body weights of the mice throughout the treatment (Fig. 1f) and conduct muscle function tests, such as forelimb and total limb grip strength tests, before and after the final treatment (Fig. 1a).
2. Follow the TREAT-NMD SOP DMD_M.2.2.001 protocol for the forelimb and total limb grip tests. Make sure the tests are blinded. Briefly, take the average of the three most consistent readings per mouse and normalize the values according to their individual body weight (Fig. 1g, h).
3. Use an animal treadmill for the run-to-exhaustion test following the protocol TREAT-NMD SOP DMD_M.2.1.003. The program is as follows: 5 m/min for 5 min, and then increase the speed by 1 m/min until exhaustion. Exhaustion is considered as the point when the mouse does not get back on the treadmill

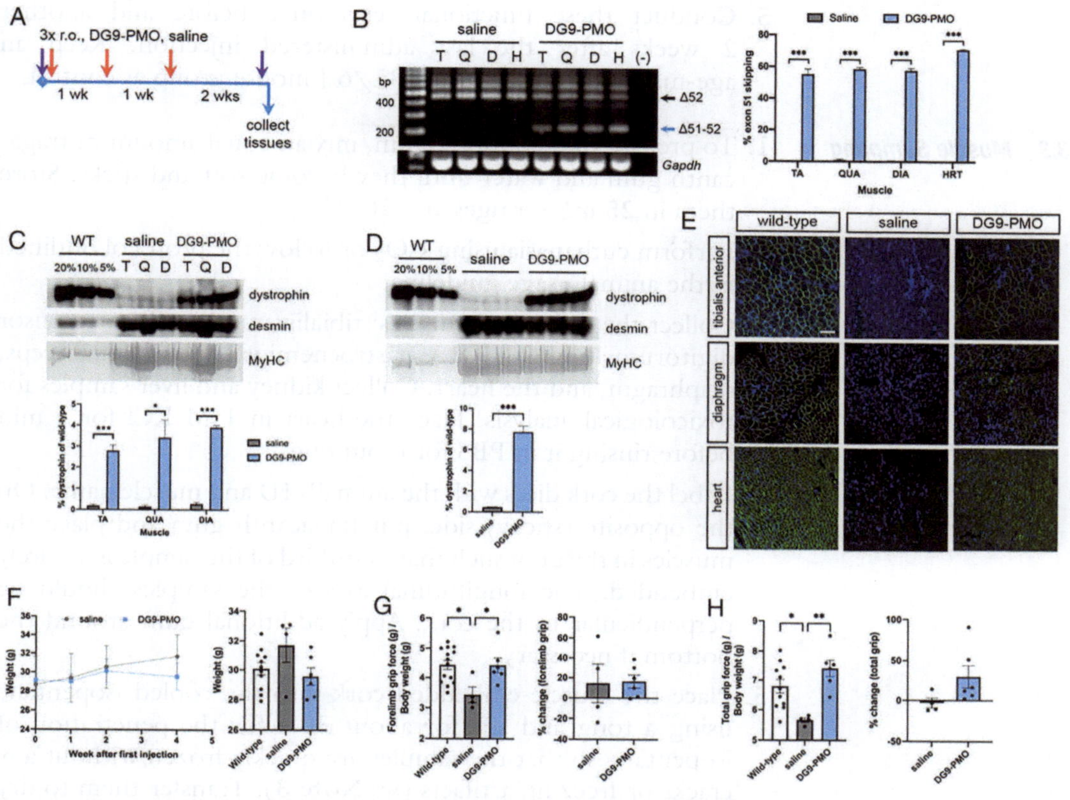

Fig. 1 Systemic administration of DG9-conjugated exon 51-skipping phosphorodiamidate morpholino oligomers (PMO). (**a**) Male, 2-month-old h*DMD*del52;mdx mice were treated three times (red arrows) via retro-orbital route with saline or 30 mg/kg DG9-PMO for exon 51 skipping, once a week for 3 weeks. Purple arrows indicate the time points for functional testing. Tissues were collected 2 weeks post-treatment. (**b**) RT-PCR showing Duchenne muscular dystrophy (*DMD*) exon 51-skipping efficiency in various muscles with quantification. *Gapdh* is shown as a control. (**c**) Western blot detection for dystrophin (DYS1) in various skeletal muscles and (**d**) in the heart. Desmin and myosin heavy chain (MyHC) serve as loading controls. Quantification is shown relative to the intensity of the 5% WT band. Error bars: S.E.M. $*p < 0.05$, $***p < 0.001$ unpaired two-tailed t-test for (**b**)–(**d**). (**e**) Immunofluorescence images for dystrophin (DYS1, green) and nuclei (DAPI, blue) in various muscles and conditions. Scale bar: 100 μm. (**f**) Changes in body weights over the course of the experiment. (**g**) Forelimb grip strength results and (**h**) total limb grip strength results normalized to body weight. Error bars: S.E.M. $*p < 0.05$, $**p < 0.005$ one-way ANOVA with Tukey's test for (**g**)–(**h**). TA/T, tibialis anterior; QUA/Q, quadriceps; DIA/D, diaphragm; HRT/H, heart. Figure reproduced from Lim et al. 2022 [5] in accordance with the CC BY-NC-ND license agreement

within 10 s following repeated, gentle nudges. Set the maximum test duration to 15 min.

4. For rotarod analysis, mice were first placed on a rod rotating at a steady speed of 5 rpm. Once all mice were in place, rotation was accelerated from 5 to 45 rpm over a span of 300 s [5]. Fall times were automatically recorded by the software. Three trials were conducted, spaced 15 min apart, and the average or peak fall time from these trials was used for analysis.

5. Conduct these functional tests once before and another 2 weeks after the last administered injection. Keep an age-matched wild-type C57BL/6 J mouse group as control.

3.3 Muscle Sampling

1. To prepare the mounting gum, mix an equal amount of tragacanth gum and water until they become soft and sticky. Store them in 25 mL syringes at −20 °C.
2. Perform euthanasia using CO_2 or follow the protocol outlined in the animal usage guideline.
3. Collect the following muscles: tibialis anterior (TA), extensor digitorum longus (EDL), gastrocnemius, soleus, quadriceps, diaphragm, and the heart. Collect kidney and liver samples for toxicological analysis. Place the heart in 1 M KCl for 1 min before rinsing it in PBS for mounting.
4. Label the cork discs with the animal's ID and muscle name. On the opposite (sticky) side, put tragacanth gum and place the muscles in the gum such that one-third of the sample is securely embedded. The longitudinal axis of the samples should be perpendicular to the cork. Apply additional gum around the bottom if necessary.
5. Place the muscle-embedded cork into the cooled isopentane using a tong and stir for about 30 s for the penetration of isopentane so that the samples are quickly frozen without any cracks or freezing artifacts (*see* **Note 3**). Transfer them to dry ice, and later store them in falcon tubes at −80 °C.

3.4 RNA Extraction and RT-PCR

1. For RNA extraction, add 1 mL of cold Trizol to the frozen sections. Then, vortex for 30 s, followed by 5 min incubation at room temperature.
2. Add 200 μL of chloroform and then vortex. Incubate for 2 min at room temperature.
3. Centrifuge at 12,000× *g* for 15 min at 4 °C and collect the supernatant in a new tube.
4. Add 1 μL of RNase-free glycogen and 500 μL of isopropanol to the solution and vortex. Incubate for 10 min at room temperature.
5. Centrifuge at 12,000× *g* for 10 min at 4 °C. RNA pellet will be at the bottom of the tube.
6. Decant the supernatant and add 1 mL of 75% ethanol.
7. Vortex briefly to dislodge the RNA pellet.
8. Centrifuge at 7500× *g* for 5 min at 4 °C.
9. Decant the supernatant and centrifuge at 7500× *g* for 5 min at 4 °C.

10. Pipette out the remaining ethanol and air-dry for 5–10 min at room temperature.
11. Add 20–30 μL of RNase-free water and warm at 65 °C for 10 min.
12. Quantify the RNA concentration using NanoDrop and add RNase-free water to get the desired concentration. Store at −80 °C.
13. For RT-PCR, prepare the samples for cDNA synthesis using the reagents as follows:
    (a) 10 mM dNTP: 1 μL
    (b) Random hexamers: 1 μL.
    (c) RNA sample: 750–1000 ng.
    (d) RNase-free water: add to make the final volume 13 μL.
14. Incubate the samples at 65 °C for 5 min, followed by 4 °C for 5 min in the PCR machine.

    Add the following reagents to the sample:
    (a) 5× RT buffer: 4 μL
    (b) 0.1 M DTT: 1 μL
    (c) RNaseOUT: 1 μL.
    (d) SuperScript IV RT: 1 μL.

    Use the following program:
    (a) 23 °C, 10 min
    (b) 52 °C, 10 min
    (c) 80 °C, 10 min
    (d) 4 °C, 5 min
    (e) 4 °C, hold.
15. For amplification, use the following reagents:
    (a) GoTaq: 12.5 μL.
    (b) 10 μm forward primer: 1.65 μL
    (c) 10 μm reverse primer: 1.65 μL
    (d) cDNA sample: 8 μL,
    (e) RNase-free water: 1.2 μL.

    Use the following program:
    (a) 95 °C, 2 min
    (b) 40 cycles of 95 °C, 30 s; 60 °C, 30 s; 72 °C, 35 s
    (c) 72 °C, 5 min
    (d) 4 °C, hold.

16. For amplification, use the following reagents:
    (a) Prepare 1.5% agarose gel and add gel red before cooling it down (1:20,000).
    (b) Run 5 μL of the PCR product at 100 V on 1.5% agarose gel and visualize.
    (c) Calculate the % of exons successfully skipped using the formula: (intensity of the skipped band/total intensity of nonskipped, intermediate, and chosen bands) × 100 (Fig. 1b).

3.5 Western Blot

1. Collect 35–40 (20 μm) sections of the muscle sample in a tube.
2. Add 80 μL of lysis buffer to the sample and vortex for 30 s.
3. Incubate at 37 °C for 5 min, followed by centrifugation at a max speed for 30 min.
4. Collect the supernatant and measure the volume.
5. Measure the protein concentration using Bradford Protein Assay Kit and adjust to the desired concentration using the lysis buffer. Store at −80 °C.
6. Mix equal parts of the protein sample with 2× LDS sample buffer:
    (a) 2.5 μL NuPAGE™ 4× LDS sample buffer
    (b) 1 μL NuPAGE™ 10× sample reducing agent
    (c) 1.5 μL water.
7. In a precast NuPAGE™ 3–8% Tris–acetate Midi gel, load 20 μL of the sample (40 μg protein) in each well.
8. Load 10 μL of the protein standard ladder.
9. Add 1× Tris–acetate SDS running buffer in both the large and small electrophoresis chambers. Add antioxidant to the running buffer in the small chamber (1:4,000).
10. Run SDS-PAGE for 75 min at 150 V.
11. Soak PVDF membrane for 20 s in methanol and then in anode buffer for 6 min.
12. Soak blotting papers in different transfer buffers for 6 min.
13. Stack the blotting papers, the membrane, and the gel in the following order:

    Anode plate: Conc. Anode–Anode–Membrane–Gel–Cathode: Cathode plate.
14. Run the semi-dry transfer at 20 V for 70 min.
15. Rinse the membrane twice with PBS.
16. Place it in 2% blocking reagent (in PBST) and shake gently for 1 h at room temperature or overnight at 4 °C for blocking.

17. Incubate the membrane with NCL-DYS1 antibody or anti-desmin antibody in a blocking solution (1:200 for NCL-DYS1 or 1:4,000 for anti-desmin) for 1 h at room temperature or overnight at 4 °C.
18. Wash the membrane three times in PBST for 10 min at room temperature.
19. Incubate the membrane with the HRP-conjugated anti-mouse IgG2a secondary antibody for DYS1 (1:10,000 in PBST), or, with the HRP-conjugated goat anti-rabbit IgG secondary antibody for desmin (1:10,000 in PBST) for 1 h at room temperature.
20. Wash the membrane three times with PBST for 10 min at room temperature.
21. Visualize using the ECL select detection kit and a chemiluminescent imager (Fig. 1c,d).

3.6 Dystrophin Immunofluorescence

1. Section samples at 7 μm thickness and place them on poly-L-lysine-coated slides. For stored samples, thaw the samples for 30 min at room temperature.
2. Prepare PBSTx (PBS with 0.1% Triton X-100) and blocking solution (10% goat serum in PBSTX).
3. Incubate the samples in the blocking solution for 2 h at room temperature.
4. Discard the blocking solution and incubate the samples overnight in a moisture chamber at 4 °C with 1:50 NCL-DYS1 (in blocking solution).
5. Wash the slides three times with PBS for 5 min each on a shaker.
6. Incubate the slides with Alexa 488-conjugated goat anti-mouse IgG2a secondary antibody (Life Technologies) at room temperature for 30 min.
7. Wash the sections three times with PBS and use DAPI-containing Vectashield HardSet Antifade Mounting Medium to mount the coverslips.
8. Visualize the samples for the presence of DAPI and dystrophin with a Zeiss LSM 710 confocal microscope at 200× magnification (Fig. 1e).

3.7 Histology

1. Section the samples of frozen muscles at 7 μm and place them on poly-L-lysine-coated slides. Store at −80 °C.
2. Before use, thaw the samples for 30 min at room temperature.

3. Place the slides in the solutions in the following order:
   (a) Mayer's hematoxylin: 15 min.
   (b) Running tap water: 15 min.
   (c) Eosin Y: 10 min.
   (d) 70% Ethanol: 5 s
   (e) 90% Ethanol: 5 s
   (f) 99% Ethanol: 30 s
   (g) 99% Ethanol: 30 s
   (h) Xylene substitute: 5 min.
   (i) Xylene substitute: 5 min.
4. Mount the muscle sections with the Permount™ (Fisher Scientific) mounting medium and gently put a coverslip above.
5. Apply nail polish around the cover slip.
6. Visualize under brightfield using a light microscope.
7. Randomly choose three fields of view for every muscle sample and calculate the CNF percentage using the formula: (# CNFs / total # fibers) × 100.
8. Count the fibers with ImageJ.
9. For every sample, obtain the average CNF % taken from every field of view.
10. Use Open-CSAM to automatically measure the minimal Feret diameters of the samples. Manually organize the measurements to ensure that the boundaries are corrected to the fibers in the sample.

3.8 Statistical Analysis

1. Perform statistical analysis using Prism v.9.0.1 (GraphPad software).
2. Conduct an unpaired two-tailed t-test or a one-way ANOVA. When appropriate, perform a post hoc Tukey's or Dunnett's multiple comparison test.

4 Notes

1. Keep the morpholino- or peptide-conjugated morpholino drugs on ice until the mouse is ready for injection. Note down the mouse weights in advance for preparing treatment dosage at a proper volume and concentration. Make sure the injection volume does not exceed the recommended volume in animal usage guidelines.
2. For our experiment, we used young adult mice that are 2 months old. The dystrophic mouse model used in the

experiments was male h*DMD*del52;*mdx* mice that have a h*DMD*del52 transgene integrated into a single allele of chromosome 5. We administered three doses of morpholino once per week for 3 weeks at 30 mg/kg body weight. The DG9 peptide (sequence N-YArVRRrGPRGYArVRRrGPRr-C; uppercase: L-amino acids, lowercase: D-amino acids) was conjugated to the PMO, Ex51_Ac0, through the 3′ end.

3. A "sludge" will appear in the bottom of the isopentane when the correct temperature is attained. Make sure to stir the samples for 20–40 s as a longer contact time causes the formation of cracks in the samples, while insufficient time can lead to freezing artifacts (TREAT-NMD SOP MDC1A_M.1.2.003).

Acknowledgments

This research was funded by The Friends of Garrett Cumming Research and Muscular Dystrophy Canada Research Chair Fund, Women, and Children's Health Research Institute (WCHRI), the University of Alberta Faculty of Medicine and Dentistry, and the University of Alberta Undergraduate Research Initiative.

Conflict of Interest The authors have no conflicts of interest to declare that are relevant to the content of this chapter.

References

1. Duan D, Goemans N, Takeda S, Mercuri E, Aartsma-Rus A (2021) Duchenne muscular dystrophy. Nat Rev Dis Primers 7:1–19
2. Nakamura A (2017) Moving towards successful exon-skipping therapy for Duchenne muscular dystrophy. J Hum Genet 62:871–876
3. Dhuri K et al (2020) Antisense oligonucleotides: an emerging area in drug discovery and development. J Clin Med 9:2004
4. Liew WKM, Kang PB (2013) Recent developments in the treatment of Duchenne muscular dystrophy and spinal muscular atrophy. Ther Adv Neurol Disord 6:147–160
5. Lim KRQ et al (2022) Development of DG9 peptide-conjugated single- and multi-exon skipping therapies for the treatment of Duchenne muscular dystrophy. Proc Natl Acad Sci 119:e2112546119
6. Koenig M et al (1989) The molecular basis for Duchenne versus Becker muscular dystrophy: correlation of severity with type of deletion. Am J Hum Genet 45:498–506
7. Shirley M (2021) Casimersen: first approval. Drugs 81:875–879
8. Lim KRQ, Maruyama R, Yokota T (2017) Eteplirsen in the treatment of Duchenne muscular dystrophy. Drug Des Devel Ther 11:533–545
9. Anwar S, Yokota T (2020) Golodirsen for Duchenne muscular dystrophy. Drugs Today (Barc) 56:491–504
10. Roshmi R, R, & Yokota, T. (2019) Viltolarsen for the treatment of Duchenne muscular dystrophy. Drugs Today 55:627
11. Tsoumpra MK et al (2019) Peptide-conjugate antisense based splice-correction for Duchenne muscular dystrophy and other neuromuscular diseases. EBioMedicine 45:630–645
12. Moulton HM, Moulton JD (2010) Morpholinos and their peptide conjugates: therapeutic promise and challenge for Duchenne muscular dystrophy. Biochimica et Biophysica Acta (BBA) – Biomembranes 1798:2296–2303
13. Politano L, Nigro G (2012) Treatment of dystrophinopathic cardiomyopathy: review of the

literature and personal results. Acta Myol 31: 24–30

14. Lim KRQ et al (2019) Efficacy of multi-exon skipping treatment in Duchenne muscular dystrophy dog model neonates. Mol Ther 27:76–86
15. Nguyen Q, Yokota T (2019) Antisense oligonucleotides for the treatment of cardiomyopathy in Duchenne muscular dystrophy. Am J Transl Res 11:1202–1218
16. Yavas A et al (2020) Detailed genetic and functional analysis of the hDMDdel52/mdx mouse model. PLoS One 15:e0244215
17. Veltrop M et al (2018) A dystrophic Duchenne mouse model for testing human antisense oligonucleotides. PLoS One 13:e0193289
18. Pastoret C, Sebille A (1995) mdx mice show progressive weakness and muscle deterioration with age. J Neurol Sci 129:97–105
19. Aslesh T, Erkut E, Ren J, Lim KRQ, Woo S, Hatlevig S, Moulton HM, Gosgnach S, Greer J, Maruyama R, Yokota T (2023) DG9-conjugated morpholino rescues phenotype in SMA mice by reaching the CNS via a subcutaneous administration. JCI Insight 8(5). https://doi.org/10.1172/jci.insight.160516
20. Shah MNA, Wilton-Clark H, Haque F, Powell B, Sutanto LE, Maradiya R, Zhabyeyev P, Roshmi RR, Anwar S, Aslesh T, Lim KRQ, Maruyama R, Bigot A, Young CS, Bittner S, Spencer MJ, Moulton HM, Oudit GY, Yokota T (2025) DG9 boosts PMO nuclear uptake and exon skipping to restore dystrophic muscle and cardiac function. Nat Commun 16(1). https://doi.org/10.1038/s41467-025-59494-8

Chapter 2

DG9-Conjugated Morpholino Rescues Phenotype in Spinal Muscular Atrophy Mice

Umme Sabrina Haque, Melissa Kohut, and Toshifumi Yokota

Abstract

Spinal muscular atrophy (SMA) is a progressive neurodegenerative disease characterized by decreased expression of the survival of motor neuron 1 (*SMN1*) gene. Antisense oligonucleotide (ASO)-based therapies have been developed recently as a means of increasing the production of functional SMN protein by targeting the *SMN2* gene. Nusinersen, an FDA-approved ASO, has shown success in SMA treatment. However, it faces several challenges that limit its therapeutic potential. One of the main challenges is the administration route, as nusinersen requires intrathecal injection, which involves direct delivery into the cerebrospinal fluid (CSF). This invasive method can be burdensome for patients and may pose accessibility and practicality issues. Another challenge is the limited ability of nusinersen to efficiently penetrate the blood–brain barrier (BBB). The BBB restricts the passage of molecules to bodywide tissues. Cost and affordability are significant challenges associated with nusinersen. The therapy is expensive, which can limit its accessibility for all patients who could benefit from it. Overcoming these challenges is crucial to optimize the therapeutic benefits of ASOs and broaden their availability to a larger population of SMA patients. To address these challenges, phosphorodiamidate morpholino oligomer (PMO)-based ASOs have emerged as a potential solution. However, PMO's inability to cross the BBB remains a significant hurdle. Recent advancements include the development of DG9 conjugated to PMO, which has demonstrated CNS distribution following subcutaneous injection. DG9-PMO treatment shows promise in increasing survival rates, improving motor coordination, mitigating muscle pathology, and enhancing neuromuscular junction characteristics, compared to the molecule equivalent to nusinersen and a benchmark cell-penetrating peptide R6G-conjugated PMO. Importantly, it also leads to a significant increase in full-length SMN2 expression in both CNS and systemic tissues. This chapter describes the methodologies employed to evaluate the efficacy and safety of DG9-PMO in a severe SMA mouse model. These methodologies include systemic ASO injection, quantitative RT-PCR, western blotting, histology, ELISA-based quantification of DG9-PMO, functional tests, and toxicology analysis. By addressing challenges related to administration route, BBB penetration, and cost, overcoming these obstacles is essential to optimize the therapeutic benefits of ASO and broaden its accessibility to a larger population of SMA patients. The advancements in DG9-PMO offer promising prospects for the development of ASO-based therapeutics.

Key words Spinal muscular atrophy (SMA), Survival motor neuron 1 (*SMN1*) gene, Antisense oligonucleotide (AON), Phosphorodiamidate morpholino oligomer (PMO), DG9-PMO, Blood, Brain barrier (BBB)

Yoshitsugu Aoki and Norio Motohashi (eds.), *Splice Switching: Methods and Protocols*, Methods in Molecular Biology, vol. 2963, https://doi.org/10.1007/978-1-0716-4738-7_2,

1 Introduction

Spinal muscular atrophy (SMA) is an autosomal recessive neurodegenerative disease that is characterized by motor neuron degeneration in the anterior horn of the spinal cord resulting in progressive muscle weakness and wasting [1–4]. This disease is caused by decreased expression of the survival of motor neuron 1 *(SMN1)* gene which encodes the survival of motor neuron (SMN) protein [1–4]. SMN protein is ubiquitously expressed and has functions in protein homeostasis and spliceosome assembly [5]. Due to intrachromosomal duplication, humans have another gene, *SMN2* [6]. Regardless, *SMN2* cannot compensate for the loss of *SMN1* due to a C-to-T transition which creates an intronic splicer silencer N1 (ISS-N1) [7, 8]. Splicing modulators can bind to ISS-N1 and exclude exon 7 in the mRNA transcript [8]. Therefore, a nonfunctional protein is produced and degraded [9]. However, around 10% of the total proteins escape the splicing modulators and are full-length functional SMN (FL-SMN) proteins (Fig. 1) [7, 9]. Concordantly, a higher number of copies of the *SMN2* gene correlates with a less severe phenotype because more FL-SMN can be produced [10]. Overall, because the *SMN2* transcript can be modified to produce the FL-SMN protein, it can be targeted as a therapeutic to increase the amount of FL-SMN and ameliorate the SMA phenotype (Fig. 1).

One therapeutic strategy is to use antisense oligonucleotides (ASOs) which enhance exon 7 inclusion in *SMN2* by binding to the ISS-N1 (Intronic Splicing Silencer N1) [11, 12]. It was previously shown that this method is effective in vitro with an 18-mer ASO

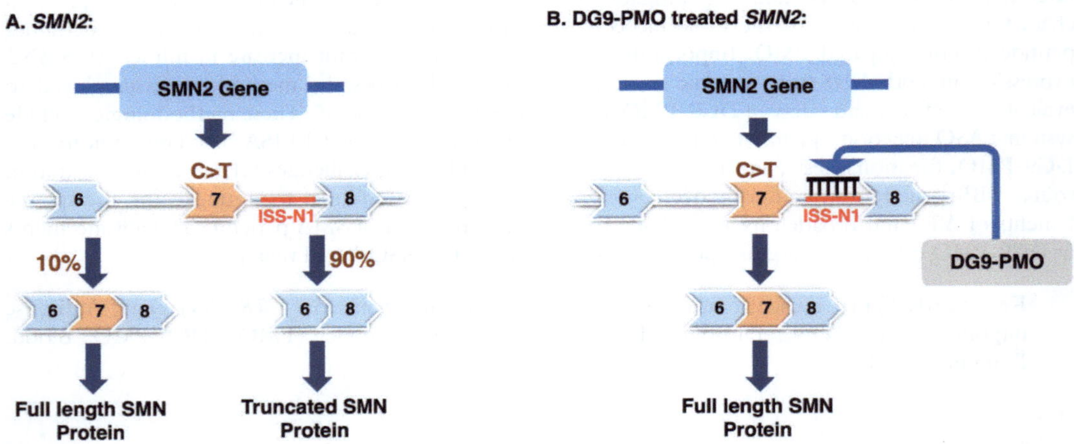

Fig. 1 Strategy to restore survival of motor neuron (SMN) expression by treatment with DG9-PMO. (**a**) Due to C-to-T transition in exon 7 of *SMN2*, it is excluded from approximately 90% of mRNA resulting in a truncated SMN protein. (**b**) After treatment with DG9-PMO, it binds to the intronic splicing silencer (ISS-N1) region of *SMN2* and produces full-length SMN protein by the inclusion of exon 7

which prevents the splicing silencers from binding [13]. One such FDA-approved ASO is nusinersen (brand name Spinraza). However, due to the impermeability of the blood–brain barrier, it must be administered repeatedly by intrathecal injection which is quite invasive and possesses a great challenge to this treatment [14]. Apart from that, due to insufficient intracellular stability, other nucleic acid chemistries have been studied for therapeutic purposes such as phosphorodiamidate morpholino oligomers (PMOs). PMOs are derivatives of deoxyribonucleotides that have an altered nucleic acid backbone of a morpholine ring in place of the pentose sugar, and the phosphodiester bonds are replaced with phosphorodiamidate linkages [11, 12]. These alterations create a neutral chemical that prevents off-target interactions with undesired cellular components, increases specificity to RNA targets, avoids degradation enzymes, and is less toxic [11]. However, PMOs face problems in reaching their RNA targets due to poor cellular uptake, rapid clearance from the circulatory system, and endosome trapping [15, 16].

To increase the efficacy of PMOs, cell-penetrating peptides (CPPs) can be conjugated to them. CPPs are derived from naturally occurring peptides that can easily penetrate the cell membranes [17]. An example of a CPP is DG9; which is derived from the human polyhomeotic 1 homolog transcription factor [18]. A study by Choi et al. showed an increase in β-galactosidase activity in various tissues including the central nervous system when the cytoplasmic domain of CTLA-4 protein was fused to the protein transduction domain of the Hph-1 transcription factor [18]. Later, a study by Aslesh et al. was conducted to prove if the DG9 protein could improve the uptake of PMOs [19]. In this study, they used the DG9 sequence derived from the protein transduction domain of Hph-1; however, the major change between these peptides is the use of D-amino acids instead of L-amino acids which decrease the toxicity [20].

Subcutaneous injection of DG9 conjugated to a PMO (DG9-PMO) improved the phenotype in the severe SMA mouse model [18]. SMA mice treated with DG9-PMO had a longer lifespan with improved muscle functioning, strength, and enhanced breathing [18]. Atrophic musculature and neuromuscular junction characteristics were both improved with the DG9-PMO treatment [18]. Most importantly, the treatment increased the expression of *FL-SMN* transcripts in both the peripheral and central nervous systems [18]. When compared to the PMO treatment only, the DG9-PMO treatment had enhanced uptake into systemic tissues and across the BBB into the central nervous system [18]. Finally, there was no observed toxicity with the treatment [18]. This chapter will discuss the techniques used to assess the effectiveness of exon skipping with DG9-PMO via subcutaneous injection into the SMA mouse model.

2 Materials

2.1 Animal Model

1. SMA-transgenic mice [FVB.Cg-Tg(SMN2)2Hung*Smn1*$^{tm1\text{-}Hung}$/J] (*Smn1*$^{-/-}$; *SMN2*$^{Tg/Tg}$) purchased from Jackson Laboratory (JAX stock 005058).
2. Heterozygous mice for Smn1 (*Smn1*$^{+/-}$; *SMN2*$^{-/-}$) were generated in the lab.
3. Litters born from breeding pairs of mice with genotype *Smn1*$^{-/-}$; *SMN2*$^{Tg/Tg}$ crossed with the heterozygous mice for Smn1 (*Smn1*$^{+/-}$; *SMN2*$^{-/-}$) consist of 50% severe type I SMA progeny genotype (*Smn1*$^{-/-}$; *SMN2*$^{Tg/-}$) (*see* **Note 1**).

The mice are kept in a specified pathogen-free environment between 70 °F and 74 °F temperature, with a 12:12 light-to-dark cycle.

2.2 SMA Genotyping

1. DNA Release (Thermo Scientific).
2. DNA Dilution Buffer (Thermo Scientific F-132 L).
3. 2X Phire Tissue Direct PCR Master Mix (Thermo Scientific).
4. Ultra-pure distilled water (Invitrogen).
5. Primer sequence for mouse *Smn* (IDT).

   Smn- S1 (5′-ATAACACCACCACTCTTACTC-3′).

   Smn-S2 (5′-GTAGCCGTGATGCCATTGTCA-3′).

   Smn-H1 (5′-AGCCTGAAGAACGAGATCAGC-3′).
6. 2% Agarose.
7. Gel red nucleic acid stain (MilliporeSigma).
8. Gene ruler 100 bp DNA ladder (Thermo Scientific).
9. Equipment: Heat block, PCR thermocycler, gel electrophoresis machine, imaging machine (ChemiDoc Imaging system, BIO-RAD), Image Lab Software.

2.3 In Vivo Delivery of ASO

1. Injection solution: PPMO powder- DG9 (sequence N-YArVRRrGPRGYArVRRrGPRr-C); hSMN_Int7_10 (sequence- 5′-TCACTTTCATAATGCTGGG-3′_18 mer) ((Gene tools LLC), 1x PBS (The C- terminus end of the peptide is covalently attached to the 3′ end of the PMO).
2. 30-gauge Hamilton syringe.
3. Phosphate-buffered saline (PBS).
4. 70% ethanol.
5. Gloves.
6. Equipment: mini centrifuge machine, vortex machine.

2.4 Harvesting, Snap-Freeze, and Cryosection of Tissue

1. Tragacanth powder (Sigma, St. Louis, MO).
2. Milli-Q water.
3. Liquid nitrogen.
4. Cork disc.
5. Isopentane.
6. Cryostat specimen disc.
7. Tissue-Tek O.C.T. compound (Sakura, Torrance, CA).
8. 25 mL syringe.
9. Equipment: Leica CM1950 Cryostat (Leica, Buffalo Grove, IL).

2.5 RNA Extraction

1. TRIzol reagent (Invitrogen, Carlsbad, CA).
2. Chloroform.
3. 100% isopropanol.
4. RNA-grade glycogen.
5. 75% ethanol.
6. RNase-free water.
7. 1.5 mL centrifuge tube (RNA grade).
8. Equipment: cooled centrifuge, heat block, NanoDrop Lite Spectrophotometer (Thermo Scientific).

2.6 cDNA Synthesis and RT-PCR

1. SuperScript IV reverse transcriptase (Thermo Fisher Scientific).
2. Oligo(dT) primers (Thermo Fisher Scientific).
3. RNaseOUT recombinant ribonuclease inhibitor (Thermo Fisher Scientific).
4. dNTP set (Thermo Fisher Scientific).
5. 2 × SSO Advance Universal SYBR Green Supermix (Bio-Rad).
6. RNase-free water.
7. Primers for amplification of the *SMN* and *GADPH*:
   (a) qPCR full-length *SMN2* forward primer (5′-GCTATCATA CTGGCTATTATATGGGTTTT-3′).
   (b) qPCR full-length *SMN2* reverse primer (5′-CTCTATGC CAGCATTTCTCCTTAAT-3′).
   (c) qPCR *SMN2Δ7* forward primer (5′-TCTGGACCACCAAT AATTCCCC-3′).
   (d) qPCR *SMN2Δ7* reverse primer (5′-ATGCCAGCATTTCCA TATAATAGCC-3′).
   (e) Total *SMN2* forward primer (5′- GCGATGATTCTGAC ATTTGG-3′).

(f) Total *SMN2* reverse primer (5′--GGAAGCTGCAGTATTCT TCT-3′).

(g) *GAPDH* forward primer (5′-GCAAATTCCATGGCACC GT-3′) (*see* **Note 6**).

(h) *GAPDH* reverse primer (5′-AGGGATCTCGCTCCTG GAA-3) (*see* **Note 6**).

8. Equipment: QuantStudio 3 (Applied Biosystems), NanoDrop 2000C spectrophotometer (Thermo Scientific, Elsmere, DE).

2.7 Protein Extraction and Western Blotting

1. Lysis buffer (10% SDS stock solution, 70 mM Tris–HCl (pH 6.7), 5 mM EDTA-2Na (pH 8.0), 5% beta-mercaptoethanol, protease inhibitor cocktail (Roche, Germany), ultrapure distilled water.
2. Pierce™ Coomassie (Bradford) protein assay kit (Thermo Scientific™, 23,200).
3. XCell4 SureLock™ Midi-Cell system (Invitrogen, WR0100).
4. Novex® semi-dry blotter (Invitrogen, SD1000).
5. NuPAGE® LDS sample buffer 4x (Invitrogen, NP0007).
6. NuPAGE® sample reducing agent 10x (Invitrogen, NP004).
7. NuPAGE Novex 4%–12% Bis–Tris Midi protein gels (Invitrogen, WG1402BOX).
8. NuPAGE® Tris–acetate SDS running buffer (20x) (Invitrogen, LA0041).
9. NuPAGE™ Antioxidant (Invitrogen, NP0005).
10. HiMark™ pre-stained protein standard (Invitrogen, LC5699).
11. Polyvinylidene difluoride (PVDF) membrane, pore size 0.45 μm (Millipore, IPFL00010).
12. Extra thick paper, 2.5 mm thickness, 8.0 cm × 13.5 cm (Thermo Fisher, PI-88615).
13. 5% skim milk.
14. Methanol.
15. Concentrated anode buffer: 0.3 M Tris, 20% methanol.
16. Anode buffer: 0.03 M Tris, 20% methanol.
17. Cathode buffer: 25 mM Tris, 20% methanol, 40 mM 6-amino-n-hexanoic acid, 0.01% SDS.
18. Phosphate-buffered saline (PBS) containing 0.05% Tween 20 (PBST).
19. Amersham ECL prime blocking reagent (GE Healthcare, Little Chalfont, UK).
20. Purified mouse anti-SMN antibody (610,647, BD Biosciences).

21. Rabbit anti-beta tubulin antibody (Abcam ab6046).
22. HRP-conjugated goat anti-mouse (IgG H + L; 31,430, Invitrogen).
23. HRP-conjugated goat anti-rabbit (IgG H + L; 31,460, Invitrogen).
24. Amersham ECL Select western blotting detection reagent (GE Healthcare).
25. Stripping buffer (for 1 L: 15 g glycine, 1 g SDS, 10 mL Tween 20, and adjust pH 2.2).
26. Equipment: mini-centrifuge machine, ChemiDoc Touch Imaging system (Bio-Rad).

2.8 ELISA (Enzyme-Linked Immunosorbent Assay)

1. Frozen tissue section.
2. Lysis buffer: 1 protease inhibitor tablet (Roche) in 7 mL of RIPA buffer (Thermo Fisher).
3. Hybridization buffer: 1x TE buffer; 1 M final NaCl solution; 0.1% v/v Triton X-100.
4. Probes complementary to the PMO sequence with digoxigenin and biotin-labeled at 5′ and 3′ ends, respectively (sequence 5′-/5DigN/C*C*A*G*C*ATTATGAAA*G*T*G*A / 3Bio/−3′) (Integrated DNA Technology).
5. Pierce NeutrAvidin-coated 96-well plates, Black (Thermo Fisher Scientific).
6. Wash buffer: 50 mM Tris–HCl, 150 mM sodium chloride pH 7.6, 0.1% v/v Tween-20.
7. Micrococcal nuclease enzyme (MNase) (New England Biolabs, cat#M0247S).
8. MNase reaction buffer: pH 7.9; 50 mM Tris–HCl, 100 mM NaCl, 5 mM CaCl2, 0.1 mg/mL bovine serum albumin.
9. Anti-digoxigenin antibody conjugated with ALP (alkaline phosphatase enzyme) (Roche Applied Sciences, cat# 11093274910).
10. AttoPhos AP fluorescent substrate (Promega).
11. Ultrapure distilled water (Sigma Aldrich).
12. Microseals (PCR Plate seal).
13. SuperBlock (TBS) blocking buffer (Thermo Fisher, cat#37581).
14. Equipment: Vortex mixer, cryostat, centrifuge, incubator, SpectraMax M3 plate reader (Molecular Devices).

2.9 Hematoxylin and Eosin (H & E) Staining

1. Hematoxylin solution, Mayer's (Sigma, St. Louis, MO).
2. Eosin Y solution (Sigma, St. Louis, MO).
3. 100% ethanol.
4. 90% ethanol.
5. 70% ethanol.
6. Distilled water.
7. Filter paper.
8. Coverslip.
9. SafeClearII Xylene substitutes (Fisher Scientific, 23–044192 or alternatives).
10. Permount Mounting Medium (Fisher Scientific, SP15100).
11. Hydrophobic barrier pen.
12. Nail polish.
13. ImageJ software.

2.10 Neuromuscular Junction (NMJ) Staining

1. 4% Paraformaldehyde (PFA).
2. Phosphate-buffered saline (PBS).
3. 5% goat serum.
4. 1% Triton X-100.
5. Primary mouse monoclonal antibodies anti-neurofilament clone 2H3 and anti-synaptophysin clone SV2 (antibodies from DSHB, Iowa).
6. Alexa Fluor 488-goat anti-mouse IgG1 (Life Technologies, A21121).
7. Alexa Fluor 594-bungarotoxin (Thermo Fisher Scientific, B13423).
8. VECTASHIELD HardSet Antifade Mounting Medium with DAPI (4′,6-diamidino-2-phenylindole dihydrochloride) (Vector Laboratories).
9. Coverslip.
10. Moisture chamber.
11. Hydrophobic barrier pen.
12. Nail polish.
13. Zeiss ZEN software.
14. Equipment: Shaker, confocal microscope (Zeiss LSM 710).

2.11 Macrophage Detection

1. 4% PFA.
2. PBS.
3. 10% goat serum.
4. 0.1% Triton X.

5. Primary antibody—rat anti-mouse CD68 (MCA 1957, BIO-RAD).
6. Secondary antibody—Alexa Fluor 488 goat anti-rat antibody (Thermo Fisher Scientific, A11006).
7. VECTASHIELD HardSet Antifade Mounting Medium with DAPI (4′,6-diamidino-2-phenylindole dihydrochloride) (Vector Laboratories).
8. Coverslip.
9. Moisture chamber.
10. Hydrophobic barrier pen.
11. Nail polish.
12. Zeiss ZEN software.
13. Equipment: Shaker, confocal microscope (Zeiss LSM 710).

2.12 Righting Reflex Test

1. Cotton sheet or bench pad.
2. Stopwatch.

2.13 Hind Limb Suspension Test

1. 50 mL conical tube.
2. Laboratory wipes.
3. Stopwatch/timer.

2.14 Forelimb Grip Strength

1. Grip strength meter: horizontal forelimb meter (Columbus Instrument, Columbus, OH).

2.15 Rotarod Test

1. AccuScan rotarod instrument.
2. Rotarod apparatus with individual lanes.
3. Stopwatch/timer.

2.16 Whole-Body Plethysmographic Recording

1. 10 mL (inner diameter: 1.9 cm, length: 3.5 cm for PD7 mice with body weight less than 2.5 g), 30 mL (inner diameter: 2.6 cm, length: 5.6 cm for PD7 mice with body weight more than 2.5 g), and 80 mL (inner diameter: 3.8 cm, length: 7 cm for PD30) volumes of Plexiglass plethysmographs with one inflow for delivery of fresh air and one outflow for removal of carbon dioxide (Validyne Engineering).
2. Gas mixer (GSM-3, CWE).
3. Gas regulators (Porter Instrument Company).
4. Gas analyzer (ML206, ADInstruments).
5. Infant incubator (Isolette, model C-86; Air-Shields/Dräger Medical).
6. Pressure transducer (DP 103; Validyne).
7. Signal conditioner (CD-15; Validyne).
8. Analog-digital board (Digidata 1322A).

9. 80% ethanol.
10. Axoscope (Data acquisition software), Sigmaplot 11 Systat Software Inc., LabChart 8 software (ADInstruments).

2.17 Urinary KIM-1 Analysis

1. KIM-1 SimpleStep ELISA® Kit (ab213477).
2. Ultrapure water.
3. 1.5 mL centrifuge tube.
4. Kimwipe.
5. Plate seal.
6. Equipment: Plate shaker, centrifuge machine, SpectraMax M3 plate reader (Molecular Devices).

3 Methods

3.1 Genotyping for Homozygous SMA Mice

1. Cut the tails (>1 mm) from neonatal pups (postnatal day 0) and put in a 1.5 mL tube.
2. Isolate genomic DNA (gDNA) using 2X Phire Tissue Direct PCR Master Mix (Thermo Scientific) (*see* **Note 2**).
   (a) Prepare the DNA extraction Master mix by adding 20 μL of dilution buffer and 0.5 μL of DNA Release buffer for each sample.
   (b) Add 20.5 μL of the extraction Master mix to each tube containing the tail samples and incubate for 5 min at room temperature.
   (c) Incubate on a heat blocker for 5 min at 98 °C and put on ice for at least 2 min.
3. Dilute the gDNA 10x and use 2 μL of diluted gDNA for PCR.
4. Make the PCR premix by adding 10 μL of Phire, 0.8 μL of 10 μM of S1, S2, and H1 primer each, and 5.6 μL of ultrapure distilled water for each sample.
5. Add 18 μL of PCR premix with 2 μL of diluted gDNA sample.
6. Amplify using the following condition(s): mouse *Smn*: 98 °C/5 min → 98 °C/5 s_58°C/10 s _72°C/10 s × 35 cycles → 72 °C/1 min.
7. Perform gel loading and image.
   (a) Add 2% agarose in 0.5% TBE buffer and microwave it.
   (b) Add 5% gel red and mix thoroughly.
   (c) Load 5 μL of the ladder and 5 μL of PCR product in each well.
   (d) Run the gel for 30 min at 100 V.
   (e) See the gel in an imaging machine to observe the bands.

3.2 Injection of PMO Drug into Mice

1. Spin down the PMO powder using a mini-centrifuge machine.
2. Make 1 mM stock solution by adding PBS and vortex to mix it.
3. Wash the syringe with 70% ethanol three times (pipette in and out a few times).
4. Wash with PBS with a plunger up and down.
5. With the thumb and index finger, gently hold the pups by the skin in the neck region.
6. Load the required volume in the syringe, wipe the injection site with ethanol, and inject the needle into the skin at a suitable angle and inject the required volume of PMO solution.
7. Pull out the needle from the skin without changing the position. Wipe the injection site with ethanol.
8. Repeat **steps 3** and **4** for washing.

3.3 Cryosection of Tissue from SMA Mice

1. Dissect the pups at postnatal day 7 and collect the following organs and tissue: left and right quadriceps, kidney, liver, intercostal muscle, diaphragm, spleen, heart, brain, and spinal cord.
2. Label the cork disc with animal ID and sample name and mount it on the cork disc using tragacanth gum (preparation of tragacanth gum: mix 20 gm of tragacanth powder with an equal volume of water and store them in a 25 mL syringe at −20 °C).
3. Quickly dip the tissue or organ-embedded cork disc into liquid nitrogen-cooled isopentane for 30 s and store it at −80 °C until use.
4. Precool the cryostat and knives to 20 °C.
5. Leave the samples in a precooled cryostat for 30 min.
6. Place the cork disc on the round metallic cryostat specimen disc and mount it with Tissue-Tek O.C.T. Leave it in the cooled cryostat chamber to get solidified.
7. Adjust the orientation to obtain a transversal section before cutting each sample throughout its length.
8. Trim the sample surface by discarding 20 sections per cut.
9. Cut a 7 μm thick section for immunostaining, H & E, and NMJ staining and 25–30 sections of 20 μm thickness for RNA extraction and protein isolation.
10. Collect the sections on slides and Eppendorf tubes for immunostaining and RNA extraction, respectively. Keep the slides at room temperature and the sample-containing Eppendorf at −20 °C.

11. After drying at room temperature, cover the slides with a cling film.
12. Store the slides and tissues at −80 °C until use.

3.4 RNA Extraction

1. After taking out the samples, immediately add 1 mL TRIzol. Mix by pipetting it up and down 10x, vortex for 30 s, and incubate at room temperature for 5 min.
2. Add 200 μL of chloroform per 1 mL TRIzol used, shake vigorously for 30 s, and incubate it at room temperature for 2–3 min (*see* **Note 3**).
3. Centrifuge the samples at 12000× *g* for 15 min at 4 °C.
4. Transfer the upper clear aqueous phase in a new RNA-grade 1.5 μL tube (*see* **Note 4**).
5. Add 1 μL of RNase-free glycogen and 500 μL 100% isopropanol per 1 mL TRIzol used. Vortex briefly to mix and incubate at room temperature for 10 min.
6. Centrifuge samples at 12000× *g* for 10 min at 4 °C.
7. A white pellet should be observed at the bottom of the tube. Discard the supernatant.
8. Add 1 mL of 75% ethanol for 1 mL of TRIzol used. Vortex briefly until the pellet is seen floating.
9. Centrifuge samples at 7500× *g* for 5 min at 4 °C.
10. Discard the supernatant. Pipette out the remaining ethanol. Air-dry samples with Kimwipe on top for 5 min at room temperature (*see* **Note 5**).
11. Resuspend the RNA pellet in 50 μL of RNase-free water and dissolve it by pipetting up and down.
12. Immediately transfer samples on ice.
13. Measure the RNA concentration using a NanoDrop Lite Spectrophotometer and store at −80 °C.

3.5 cDNA Synthesis and RT-qPCR

1. Synthesize cDNA using SuperScript IV reverse transcriptase.
   (a) Thaw reagents and RNA on ice. Prepare the following reaction mix by adding 1 μL oligo dT, 1 μL of 10 mM dNTP, and 11 μL of 50 ng/μL RNA template. Vortex briefly and incubate at 65 °C for 5 min and then at 4 °C for 1 min:
   (b) Add 4 μL of 5× first-strand buffer, 1 μL of 0.1 M DTT, 1 μL of 40 U/μL RNaseOUT, 200 U/ μL of SuperScript IV reverse transcriptase. Mix by pipetting gently up and down.

(c) Incubate the reaction mix at 50 °C for 10 min.
(d) Inactivate the reaction by heating at 80 °C for 10 min.
(e) Dilute the cDNA product at a 1:5 ratio with DNase-free water and store at −20 °C.

2. Perform qPCR using 2x SSO Advance Universal SYBR Green Supermix. Thaw the reagents on ice.
3. Three separate reactions were carried out for three different primer sets (full-length *SMN2*, *SMN2Δ7, and* total *SMN2)*.
4. For Master mix, add 10 μL of 2 x SSO Advance Universal SYBR Green Supermix, 0.8 μL of 10 μM respective forward and reverse primers, 5.4 μL of ultrapure DNase-free water, 3 μL of diluted cDNA template (total volume for each reaction mixture is 20 μL).
5. Use an equal volume of water instead of cDNA for negative control.
6. Vortex the reaction mixture briefly.
7. Amplify using following conditions: 95 °C for 30 s → 95 °C/5 s_60°C/20 s × 35 cycles.
8. Calculate the relative expression using the $\Delta\Delta C_t$ method.
9. Plot the graph.

3.6 Protein Extraction and Western Blotting

1. Take out the samples from −80 °C, place them on dry ice, and isolate the protein according to the steps below:
(a) Add 80 μL of lysis buffer and vortex for 30 s.
(b) Spin down the samples using a mini-centrifuge.
(c) Incubate the samples at 37 °C for 5 min to denature the protein.
(d) Centrifuge the samples at max speed for 30 min at 16 °C.
(e) Collect the supernatant in a new 1.5 mL centrifuge tube and measure the volume.
(f) Dilute the extracted protein to 50-fold with ultrapure distilled water.
2. Quantify protein concentration using the Bradford Assay Kit and dilute the samples with lysis buffer to the concentration of 0.5 μg/μL.
3. Make 2x LDS sample buffer using 4× LDS loading buffer, 10x reducing agent, and water at a ratio of 2.5:1:1.5.
4. Mix 12 μL of diluted protein with 12 μL of 2× LDS sample buffer. Heat the mixed samples at 70 °C for 10 min and spin them down.

5. In a graduated cylinder, add 760 mL of milliQ water and add 40 mL of NuPage SDS running buffer and mix well. Then, transfer 200 mL from the mixture into a smaller bottle and add 500 μL of antioxidant to it for use in the upper buffer chamber of the SDS-PAGE tank for running one gel in the tank. For running two gels, transfer 400 mL from the mixture into a bottle and add 1000 μL of antioxidant to it for use in the upper buffer chamber (*see* **Note 8**).
6. Assemble the XCell4 SureLock™ Midi-Cell system with NuPAGE Novex 4%–12% Bis–Tris Midi protein gels according to the manufacturer's guidelines and slowly load the 1x running buffer containing the antioxidant reagent into the upper buffer chamber and completely cover the sample wells.
7. Load 20 μL of the sample (5 μg protein) per well of the gel with 10 μL of protein ladder and run SDS-PAGE at 150 V for 60 min.
8. Equilibrate the gel after SDS-PAGE with cathode buffer for 5–10 min under agitation.
9. Soak the PVDF membrane for 20 s in methanol under agitation and then in the anode buffer for at least 10 min.
10. Soak blotting papers in each transfer buffer for at least 30 min.
11. Place the blotting papers, the PVDF membrane, and the gel on a semi-dry transfer machine, as shown in Fig. 2 (*see* **Note 9**).
12. Transfer at 20 V of constant voltage for 30 min.
13. Rinse the membrane with PBST (0.05% Tween 20 in PBS) to remove the transfer buffer completely.
14. Block the membrane in 5% skim milk in PBST and shake gently overnight at 4 °C.
15. Incubate the membrane with purified mouse anti-SMN antibody in a blocking solution (1:10,000) for 1 h at room temperature under agitation.

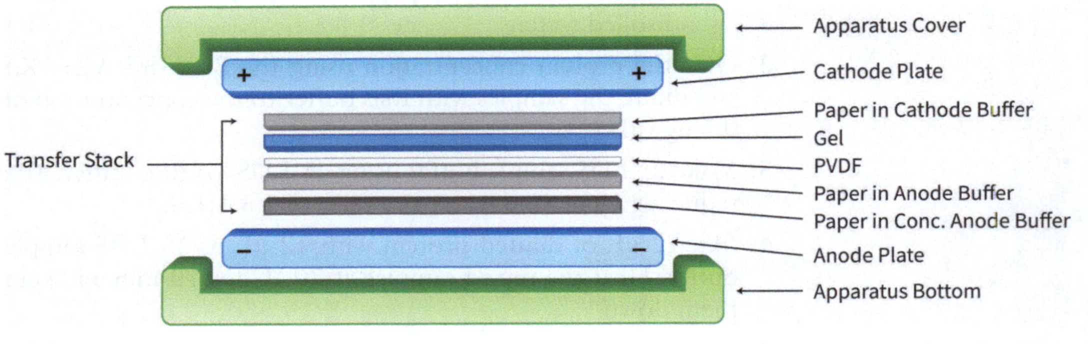

Fig. 2 Semi-dry transfer method for western blotting. The order presented in the figure is blotting paper in conc. Anode buffer, blotting paper in anode buffer, polyvinylidene difluoride (PVDF) membrane soaked in anode, SDS-PAGE gel soaked in cathode buffer, and blotting paper soaked in cathode buffer

16. Wash the membrane three times for 10 min by PBST at room temperature.
17. Incubate the membrane with the HRP-conjugated goat anti-mouse in PBST (1:10,000) for 1 h at room temperature under agitation.
18. Wash the membrane three times for 10 min by PBST at room temperature.
19. The bands are detected using the Amersham ECL Select Western Blotting Detection Kit (GE Healthcare) and visualized by the ChemiDoc Touch Imaging system (Bio-Rad).
20. For the detection of β-tubulin as a loading control protein, the membrane is incubated with a stripping buffer for 10 min at room temperature.
21. Wash the membrane two times for 10 min with PBS at room temperature.
22. Wash the membrane two times for 5 min with Tris-buffered saline with 0.05% Tween 20 at room temperature.
23. Block the membrane in 5% skim milk in PBST and shake gently overnight at 4 °C.
24. Incubate the membrane with rabbit anti-beta tubulin antibody in a blocking solution (1:5000) for 1 h at room temperature under agitation.
25. Wash the membrane three times for 10 min by PBST at room temperature.
26. Incubate the membrane with the HRP-conjugated goat anti-rabbit in PBST (1:10,000) for 1 h at room temperature under agitation.
27. Wash the membrane three times for 10 min by PBST at room temperature.
28. The bands are detected using the Amersham ECL Select Western Blotting Detection Kit (GE Healthcare) and visualized by the ChemiDoc Touch Imaging system (Bio-Rad).
29. For comparative analysis, compare the band intensity of SMN with regard to β-tubulin in each well.

3.7 ELISA (Enzyme-Linked Immunosorbent Assay)

1. Prepare a tissue sample from an AO-treated animal.
   (a) Collect 30–40 sections of 20 μM of tissue samples in a 1.5 μL tube. Add 150 μL of lysis buffer and vortex until sections are jelly-like.
   (b) Incubate the mixture at 55 °C overnight.
   (c) Centrifuge the samples at max speed for 15 min and collect the supernatant.

(d) Quantify the protein concentration and adjust it to 2 mg/mL using the lysis buffer.

(e) Add 22 μL of the protein lysate to 198 μL ultrapure distilled water.

(f) Store at −20 °C until use.

2. Prepare the PPMO standard.

(a) Prepare 1 μM and serially dilute to 10 nM and 5 nM stock:

For 1 μM stock, add 1 μL of 1 mM PPMO in 1498.5 μL distilled water.

For 10 nM stock, add 5 μL of 1 μM PPMO in 495 μL distilled water.

For 5 nM stock, add 500 μL of 10 nM PPMO in 30 μL lysis buffer and 470 μL distilled water.

(b) Prepare the solvent by combining 2.7 mL distilled water with 300 μL RIPA buffer. Serially dilute the PPMOs using a solvent to the following concentrations: 1400, 1200, 1000, 800, 600, 400, 200, and 0 pM. Start with the 5 nM stock: add 61.6 μL PPMO in 158.4 μL solvent to get 1400 pM concentrations and so on. Transfer 220 μL of each standard into a fresh tube.

3. Prepare the probe.

(a) Make a 100 nM stock by combining 1.5 μL of 100 μM probe with 1498.5 μL hybridization buffer.

(b) Add 100 μL of the 100 nM stock with 19.9 mL hybridization buffer to make a 0.5 nM working solution.

4. Add 200 μL of 0.5 nM probe to the standard PPMO and sample tubes.

5. Incubate the tubes for 30 min at 37 °C.

6. Wash the Avidin-coated plate three times with 200 μL wash buffer (*see* **Note 10**).

7. Transfer 200 μL of the mixture from step 4 to each well in the plate. Do this in duplicates.

8. Incubate for 30 min at 37 °C.

9. Wash the plates three times with 200 μL wash buffer (*see* **Note 10**).

10. Prepare a working concentration of 0.1 U/μL by adding 1.5 μL MNase with 30 mL reaction buffer.

11. Add 200 μL 0.1 U/μL MNase solution to each well.

12. Incubate for 1 h at 37 °C.

13. Wash the plates three times with 200 μL wash buffer (*see* **Note 10**).

14. Prepare 1:5000 anti-digoxigenin antibody in SuperBlock buffer with 0.25% Tween 20. Add 200 μL to each well followed by incubation for 1 h at 37 °C.
15. Wash the plates three times with 200 μL wash buffer (*see* **Note 10**).
16. Add 150 μL AttoPhos substrate per well and incubate for 1 h at 37 °C (*see* **Note 11**).
17. Detect fluorescence at 444 nm excitation and 555 nm emission by using a monochromator SpectraMax M3 plate reader.

3.8 Hematoxylin and Eosin (H & E) Staining

1. Take out the slides (cryosections of the quadriceps muscle, diaphragm, and intercostal muscle) from −80 °C and leave them at room temperature for 10 min.
2. Make a border around the sample on the slide with a hydrophobic pen.
3. Stain in Mayer's hematoxylin for 10–15 min (filter hematoxylin with a filter paper before use). Gently tap the slide tray on a dry napkin to remove excess liquid.
4. Wash the slides in a bucket with running tap water for 15 min (do not directly flow water on the slides).
5. Stain in Eosin for 10 min. Gently tap the slide tray on a dry napkin to remove excess liquid.
6. Dehydrate slides through 70% and 90% ethanol for 10 s each and later through 99% ethanol twice for 30 s each.
7. Soak slides in xylene substitutes I and II for 5 min each, respectively.
8. Apply 1–2 drops of Permount Mounting Medium on slides and mount with cover slips avoiding air bubbles. Secure the cover slip around the edges with nail polish.
9. Acquire the respective digital images of each slide and quantify CSA and minimal Feret's diameter using ImageJ (NIH).

3.9 Neuromuscular Junction (NMJ) Staining

1. Take out the samples (cryosections of quadriceps and intercostal muscles) and leave them at room temperature for 30 min.
2. Fix the samples in 4% paraformaldehyde for 2 h on ice.
3. Wash the slides thoroughly with PBS on a shaker three times, 5 min each. Keep the slides covered during this step.
4. Prepare the blocking reagent by adding 5% goat serum and 1% Triton X-100 in PBS. Add the blocking agent to the sample and incubate for 1 h at room temperature.
5. Dilute the primary antibodies (anti-neurofilament and anti-synaptophysin) in the blocking reagent (1:500). Add muscle cryosections and incubate at 4 °C overnight in a moisture chamber.

6. Quickly wash with PBS three times on a shaker. Keep the slides covered.
7. Dilute Alexa Fluor 488 goat anti-mouse IgG (1:1000) and Alexa Fluor 594- bungarotoxin (1:5000) in PBS. Add the samples and incubate in a moisture chamber for 4 h at room temperature.
8. Wash thoroughly with PBS three times on a shaker, 5 min each. Keep the slides covered.
9. Wipe off excess liquid. Add DAPI containing mounting medium on samples.
10. Place the coverslips on the sample avoiding any air bubbles. Let the slides air-dry, and seal the slides using a nail polish.
11. To acquire high-resolution z-stack project images, a confocal microscope is used.
12. At least 300 NMJs were stacked to analyze the morphology of denervation/innervation and CSA using Zeiss ZEN software.

3.10 Macrophage Detection

1. Take out the slides and air-dry them for 30 min at room temperature.
2. Draw a border enclosing the muscle section on the slide with a hydrophobic pen.
3. Incubate the slides in 4% PFA for 5 min at room temperature.
4. Wash the slides in PBS three times for 5 min each on the shaker. Keep the slides covered during this step.
5. Prepare the blocking reagent by adding 10% goat serum and 0.1% Triton X-100 in PBS. Add the blocking agent to the sample and incubate for 20 min at room temperature.
6. Dilute rat anti-mouse CD68 primary antibody at a 1:200 ratio in blocking solution, add the sample, and incubate in a moisture chamber overnight at 4 °C.
7. Wash three times with PBS for 5 min each on a shaker. Keep the slides covered during washing.
8. Incubate in PBS with the secondary antibody—Alexa fluor 488 goat anti-rat antibody (1:1000 ratio).
9. Wash in PBS three times for 5 min each on a shaker. Keep the slides covered during this step.
10. Wipe off excess liquid without touching the sections on the slides. Add DAPI containing mounting medium on the samples.
11. Place a coverslip on the samples avoiding any air bubbles. Let the slides dry for 15 min at room temperature. Seal the slides by applying nail polish around the edges of the coverslip.
12. Visualize the slides under a confocal microscope and analyze them.

3.11 Righting Reflex Test

1. Place the neonatal pups (tested from PD2-PD20) on their back on a cotton sheet or bench pad and hold them in position.
2. Release the pups and record the time it takes to reposition by placing all four paws on the ground.
3. Repeat for three trials with at least 10 min resting period between each trial.
4. Each trial is done for 1 min if needed.
5. The data are expressed as mean ± standard error of the mean (SEM).

3.12 Hindlimb Suspension Test

1. Prepare the conical tube by putting the laboratory wipes at the bottom of the tube.
2. Place the pups in such a way so that the face is down into the tube and the hind limb suspends outside from the tube (tested for PD2- PD12).
3. Release the pups. Score based on the positioning of their legs according to the following criteria and record the latency of fall using a stopwatch.
   (a) Score 4: if the hindlimb is separated normally and the tail is raised.
   (b) Score 3: if the hindlimb is close to each other with seldom touch.
   (c) Score 2: if the hindlimb is very close to each other and often touches.
   (d) Score 1: if hindlimbs are almost always in a clasped position.
   (e) Score 0: if the pups fail to hold on to the tube for any time period.
4. Repeat the trial three times with at least 15 min of resting period. The cutoff time for this test is 30 s.

3.13 Forelimb Grip Strength

1. Set the meter of the instrument at zero. Choose kilogram-force (KFG) as a unit of value for the forelimb grip strength meter.
2. Gently pick up a mouse and allow it to grasp the grid or bar with its forelimb and lift the mouse by the tail horizontally (*see* **Note 12**).
3. Make sure the mouse hindlimb is not in contact with the grid during the test.
4. Slowly and steadily pull the mouse with its tail horizontally away from the grid until it releases the grip.
5. Record the number of grip strength shown on the grip strength meter.

6. Repeat this three times for each mouse with 15 min of resting period.
7. PD30 and PD60 mice were used for this test.

3.14 Rotarod Test

1. Power on the AccuScan rotarod and allow it to initialize.
2. Gently pick up the mouse and place it on the rotating rod of an individual lane on the rotarod apparatus.
3. Set the speed to 4 RPM, which rises to 40 RPM after 300 s (e.g., an increase of 1 RPM is achieved approximately every 8.3 s).
4. Each mouse goes through three trials. Each trial ended when the mouse slipped off the rod.
5. The latency of time is recorded.
6. Calculate the average time on the rotarod for each mouse by averaging the recorded value.

3.15 Whole-Body Plethysmographic Recording

1. Connect the data acquisition machine to the laptop/desktop. Turn it on and open the data acquisition software. The software will automatically detect the equipment setup.
2. Record the temperature, humidity, and barometric pressure of the plethysmographic chamber before placing the mouse in it.
3. After weighing, place the mouse in the chamber and cover the open end slightly for acclimatization. Close the chamber (*see* **Note 13**).
4. Start the flow of the gas. Gas was mixed in a gas chamber, delivered by pure oxygen and nitrogen canister, and monitored using a gas regulator.
5. For measuring the respiratory parameter, set the flow rate of 15, 45, and 120 mL/min for PD7 mice with body weight less than 2.5 g, PD7 mice with body weight more than 2.5 g, and PD30 mice, respectively.
6. By switching the inflow gas from normoxia (21% oxygen, balanced by nitrogen) to hypoxia (11% of oxygen for 5 min), a hypoxic challenge was carried out with continuous monitoring of plethysmographic recordings without handling the mouse physically.
7. The gas exchange took 1 min, and it was confirmed using a gas analyzer.
8. Start the recording on the software program. For PD30, the recording was done at 22 °C, and for PD7, the plethysmograph was placed in an infant incubator to maintain the nest temperature of 32 °C (*see* **Notes 14** and **15**).

9. Stop the recording after 1 min. Immediately record the chamber temperature and pressure change using a pressure transducer.
10. Record the signal conditioner with an Axoscope using an analog-digital board. The signals were high-pass-filtered (0.01 kHz) with a 1 kHz sampling rate.
11. Return the mouse to its respective cage and clean the chamber with 80% ethanol.
12. Before starting the procedure for another mouse, allow the chamber to return to the baseline temperature and humidity.
13. Plethysmography analysis and result representation:
    (a) Analyze the changes relative to the wild-type normoxia by using a plethysmograph that provided semiquantitative measurement of V_T (tidal volume) and minute ventilation ($V_E = f_R \times V_T$: mL/min/g); f_R = frequency of breathing.
    (b) CV (measure of relative variability) is a ratio of the standard deviation to the mean. A smaller CV indicates more regular breathing.
    (c) Respiratory parameters V_T and V_E were reported relative to the mean of heterozygotes in normoxia (100%).
    (d) First, run the normality test and equal variance test by using Shapiro–Wilk and Brown–Forsythe tests, respectively.
    (e) The tests that pass both normality and variance test, and parametric statistics, were used with two-way repeated measures of ANOVA, followed by the Holm–Šidák method (two factors: different treatments and different conditions).
    (f) Nonparametric tests were applied for those tests that did not pass any of the tests.
    (g) Comparison of the difference in normoxia or hypoxia was conducted with the Kruskal–Wallis one-way ANOVA on ranks, followed by Dunn's method.
    (h) A signed-rank test was used to determine how hypoxia and normoxia differed from one another.
    (i) To examine the correlation between respiratory phenotype and a decrease in body weight, Pearson product–moment correlation *t* was carried out.
    (j) Sigmaplot 11 was used to express whole-body plethysmography recording data as mean ± SD, or first interquartile 25%, median 50%, and third interquartile 75%.

3.16 Urinary KIM-1 Analysis

1. For the preparation of sample for KIM-1 assay, collect urine from PD30 mice in a 1.5 mL centrifuge tube, centrifuge at 2000 *rcf* for 10 min at 4 °C, followed by collection of the supernatant in a fresh 1.5 mL tube.
2. Prepare standard for KIM-1 assay.
   (a) Reconstitute the KIM-1 protein standard by adding 1 mL sample diluent NS. Incubate at room temperature for 10 min and mix thoroughly. The concentration of the stock standard solution is 670 pg/mL.
   (b) Label eight 1.5 mL tubes. Add 121 μL sample diluent NS into tube number 1 and 150 μL into tubes 2–8.
   (c) For serial dilution, add 179 μL of the stock standard from step 2a to tube number 1 and mix thoroughly. Take 150 μL of solution from tube 1 and add it to tube 2 and so on, as shown in Fig. 3. By this, tube 8 contains no protein and is the blank control (*see* **Note 18**).
3. Add 50 μL standard or sample to their respective well.
4. Add 50 μL of the antibody cocktail with the sample or standard.
5. Seal the plate with a plate seal and incubate on a plate shaker at 400 rpm for 1 h at room temperature.
6. Wash each well with 350 μL 1X wash buffer PT three times by decanting from wells and then dispensing 350 μL 1X wash buffer PT into each well. After the final wash, remove the excess solution by inverting the plate against Kimwipe (*see* **Note 19**).
7. Add 100 μL of TMB substrate into each well and incubate at room temperature on a shaker (400 rpm) for 10 min in the dark.

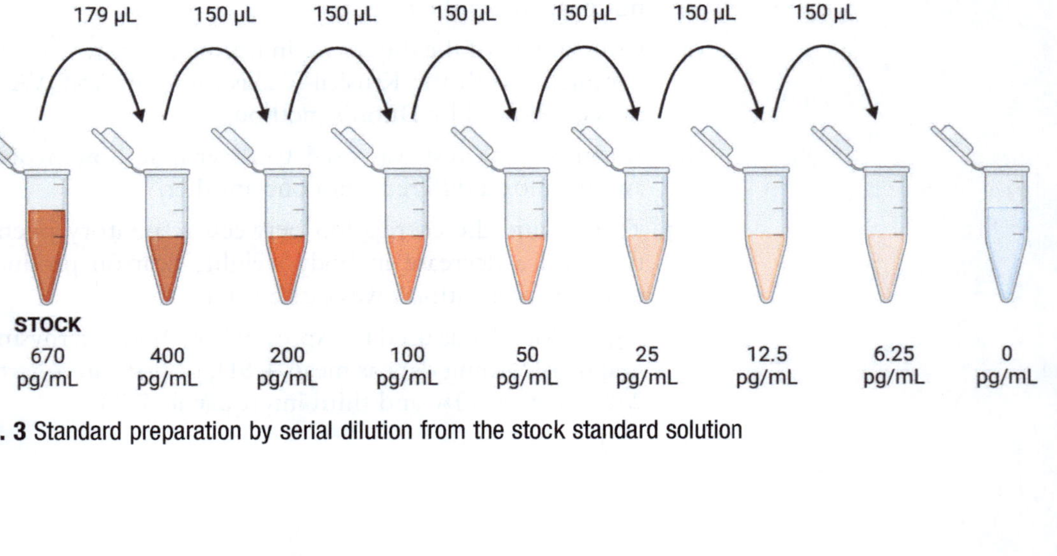

Fig. 3 Standard preparation by serial dilution from the stock standard solution

8. Add 100 μL of stop solution to each well and mix thoroughly by shaking the plates on a shaker for 1 min.
9. Record the optical density at 450 nm by a SpectraMax M3 plate reader (*see* **Note 20**).

4 Notes

1. Taiwanese male mice take a longer time to reproduce than wild-type due to their impaired reproductive organ development and reduced fertility [21]. Therefore, it is advised to breed male heterozygous wild-type mice with female type III SMA mice to create the severely affected type I SMA mice $Smn^{-/-}$; $SMN2^{Tg/-}$. It is advised that breeders be updated every 6 months, especially for breeders with type III SMA. Female SMA mice should not reproduce for the first time until they are 12 weeks old.
2. Make sure to vortex and spin down the reagents before using. However, do not vortex the enzyme as vigorous vortex can degrade the enzyme.
3. For RNA extraction, we advise adding chloroform in all the samples together and shaking the tubes at once.
4. While transferring the aqueous phase, it is suggested not to touch the colored layer to avoid contamination.
5. It is advised to pipette out ethanol carefully; otherwise, there is a chance of dislodging the pellet and losing while discarding the ethanol.
6. During RT-PCR, *GAPDH* is used as an endogenous control.
7. Usually, RNAase contamination is the primary cause of RT-PCR failure. To prevent RNAase contamination, the entire RT-PCR technique, from material and reagent preparation to RNA extraction, needs to be meticulously followed.
8. Ensure that the upper buffer chambers are not leaking.
9. It is advised to handle the PVDF membrane with plastic forceps. Remove any air bubbles with a blotting roller.
10. Avoid pipetting out the wash buffer. Remove it by quickly and continuously turning the plate over.
11. When using a 96-well plate, add 4 μL of anti-digoxigenin antibody in 20 mL of SuperBlock blocking buffer with 0.25% Tween 20.
12. For the grip strength test, the mouse should be allowed to fully grasp the meter mesh in the proper direction throughout the test and should always be dragged in the same direction by the same amount of force to minimize variation.

13. For whole-body plethymography recording, it is advised to introduce the mouse to the plethysmographic chamber to acclimatize and reduce stress 1 week before the experiment.
14. Movement will alter the animal's basal respiratory physiology and provide poor results.
15. In case of urination and defecation during recording, stop recording immediately, and clean the chamber with 80% ethanol; otherwise, it will give obscure results.
16. Plethysmography is effective for studying f_R and detection of apneas (absence of airflow (pressure changes) for a period equivalent to or greater than two complete respiratory cycles).
17. For KIM-1 urinary analysis, it is recommended to assay all standards, controls, and samples in duplicate.
18. Serially diluted standards and a fresh set of positive controls should be prepared immediately before use.
19. Complete removal of the liquid at each step is essential for good performance.
20. We also advise evaluating the following set of toxicity markers after treating with DG9-PMO to test liver and kidney functions—alkaline phosphatase (ALP), alanine transaminase (ALT), aspartate transferase (AST), gamma-glutamyl transpeptidase (GGT), blood urea nitrogen, creatinine, creatine kinase (CK), glucose, globulin, albumin, total bilirubin, and total protein. The range of the toxicity markers can be identified by performing ELISA (as mentioned in Subheading 3.7) using the respective antibodies.

Acknowledgments

This work was supported by the University of Alberta Faculty of Medicine and Dentistry, Slipchuk SMA Research Foundation Research Grant, the Canadian Institutes of Health Research (CIHR), the Friends of Garrett Cumming Research Funds, HM Toupin Neurological Science Research Funds, the Muscular Dystrophy Canada, the Canada Foundation for Innovation, Alberta Enterprise and Advanced Education, and the Women and Children's Health Research Institute (WCHRI). We would also like to thank Tejal Aslesh for her kind assistance.

Conflict of Interest The authors have no conflicts of interest to declare that are relevant to the content of this chapter.

References

1. Nance JR (2020) Spinal muscular atrophy. Continuum (Minneap Minn) 26:1348–1368. https://doi.org/10.1212/CON.0000000000000918
2. Mercuri E, Pera MC, Scoto M et al (2020) Spinal muscular atrophy — insights and challenges in the treatment era. Nat Rev Neurol 16: 706–715. https://doi.org/10.1038/s41582-020-00413-4
3. Reilly A, Chehade L, Kothary R (2023) Curing SMA: are we there yet? Gene Ther 30: 8–17. https://doi.org/10.1038/s41434-022-00349-y
4. Wirth B, Karakaya M, Kye MJ et al (2020) Twenty-five years of spinal muscular atrophy research: from phenotype to genotype to therapy, and what comes next. Annu Rev Genom Hum Genet 21:231–261. https://doi.org/10.1146/annurev-genom-102319
5. Chaytow H, Huang YT, Gillingwater TH et al (2018) The role of survival motor neuron protein (SMN) in protein homeostasis. Cell Mol Life Sci 75:3877–3894. https://doi.org/10.1007/s00018-018-2849-1
6. Rochette CF, Gilbert N, Simard LR (2001) SMN gene duplication and the emergence of the SMN2 gene occurred in distinct hominids: SMN2 is unique to Homo sapiens. Hum Genet 108:255–266. https://doi.org/10.1007/s004390100473
7. Lorson CL, Hahnen E, Androphy EJ et al (1999) A single nucleotide in the *SMN* gene regulates splicing and is responsible for spinal muscular atrophy. PNAS 96:6307–6311. https://doi.org/10.1073/pnas.96.11.6307
8. Singh NK, Singh NN, Androphy EJ et al (2006) Splicing of a critical exon of human survival motor neuron is regulated by a unique silencer element located in the last intron. Mol Cell Biol 26:1333–1346. https://doi.org/10.1128/mcb.26.4.1333-1346.2006
9. Vitte J, Fassier C, Tiziano FD et al (2007) Refined characterization of the expression and stability of the SMN gene products. Am J Pathol 171:1269–1280. https://doi.org/10.2353/ajpath.2007.070399
10. Feldkötter M, Schwarzer V, Wirth R et al (2002) Quantitative analyses of *SMN1* and *SMN2* based on real-time LightCycler PCR: fast and highly reliable carrier testing and prediction of severity of spinal muscular atrophy. Am J Hum Genet 70:358–368. https://doi.org/10.1086/338627
11. Amantana A, Iversen PL (2005) Pharmacokinetics and biodistribution of phosphorodiamidate morpholino antisense oligomers. Curr Opin Pharmacol 5:550–555. https://doi.org/10.1016/j.coph.2005.07.001
12. Sheng L, Rigo F, Bennett CF et al (2020) Comparison of the efficacy of MOE and PMO modifications of systemic antisense oligonucleotides in a severe SMA mouse model. Nucleic Acids Res 48:2853–2865. https://doi.org/10.1093/nar/gkaa126
13. Touznik A, Maruyama R, Hosoki K et al (2017) LNA/DNA mixmer-based antisense oligonucleotides correct alternative splicing of the SMN2 gene and restore SMN protein expression in type 1 SMA fibroblasts. Sci Rep 7(3672). https://doi.org/10.1038/s41598-017-03850-2
14. Hoy SM (2017) Nusinersen: First global approval. Drugs 77:473–479. https://doi.org/10.1007/s40265-017-0711-7
15. Lebleu B, Moulton HM, Abes R et al (2008) Cell penetrating peptide conjugates of steric block oligonucleotides. Adv Drug Deliv Rev 60:517–529. https://doi.org/10.1016/j.addr.2007.09.002
16. Tsoumpra MK, Fukumoto S, Matsumoto T et al (2019) Peptide-conjugate antisense based splice-correction for Duchenne muscular dystrophy and other neuromuscular diseases. EBioMedicine 45:630–645. https://doi.org/10.1016/j.ebiom.2019.06.036
17. van den Berg A, Dowdy SF (2011) Protein transduction domain delivery of therapeutic macromolecules. Curr Opin Biotechnol 22: 888–893. https://doi.org/10.1016/j.copbio.2011.03.008
18. Choi JM, Ahn MH, Chae WJ et al (2006) Intranasal delivery of the cytoplasmic domain of CTLA-4 using a novel protein transduction domain prevents allergic inflammation. Nat Med 12:574–579. https://doi.org/10.1038/nm1385
19. Aslesh T, Erkut E, Ren J et al (2023) DG9-conjugated morpholino rescues phenotype in SMA mice by reaching the CNS via a subcutaneous administration. JCI Insight 8: e160516. https://doi.org/10.1172/jci.insight.160516
20. Wu RP, Youngblood DS, Hassinger JN et al (2007) Cell-penetrating peptides as transporters for morpholino oligomers: effects of amino acid composition on intracellular delivery and

cytotoxicity. Nucleic Acids Res 35:5182–5191. https://doi.org/10.1093/nar/gkm478

21. Ottesen EW, Howell MD, Singh NN et al (2016) Severe impairment of male reproductive organ development in a low SMN expressing mouse model of spinal muscular atrophy. Sci Rep 6:20193. https://doi.org/10.1038/srep20193

Chapter 3

Cardio-Physiological Evaluations in a Dog Model with Duchenne Muscular Dystrophy

Koichi Kimura and Hiroyuki Morita

Abstract

Evaluation of cardiac function is critically important in animal experiments, particularly using disease models with cardiac complications. Here, we describe detailed methods and representative baseline results of cardio-physiological evaluations for a dog model with Duchenne muscular dystrophy (CXMDj).

Key words Animal model, Dog, Duchenne muscular dystrophy, CXMDj, Cardiac evaluation, Echocardiography, Electrocardiogram, Magnetic resonance imaging

1 Introduction

Duchenne muscular dystrophy (DMD) is an X-linked recessive myopathy caused by *dystrophin* mutations. The juvenile patients suffer from inevitable skeletal muscle weakness and progressive cardiomyopathy. In recent years, several oligonucleotide drugs, which have antisense exon-skipping effects, have been approved as treatment drugs for DMD in the United States and Japan. A dog model of DMD, canine X-linked muscular dystrophy in Japan (CXMDj), has been bred in National Center of Neurology and Psychiatry (NCNP) and used for many preclinical studies including antisense splicing modulation, morpholino exon skipping, and other CRISPR gene therapy experiments (Fig. 1). In this section, we describe the detailed methods and representative baseline results of cardio-physiological evaluations, which have been performed in NCNP. Many figures to visualize practical methods and several tips with baseline result data would be of great help for the readers to refer.

In NCNP, cardio-physiological evaluations for CXMDj are performed under isoflurane anesthesia, and the evaluation methods include echocardiography, 12-lead electrocardiogram (ECG), 24-h Holter ECG, and cardiac magnetic resonance imaging (MRI).

Yoshitsugu Aoki and Norio Motohashi (eds.), *Splice Switching: Methods and Protocols*, Methods in Molecular Biology, vol. 2963, https://doi.org/10.1007/978-1-0716-4738-7_3,

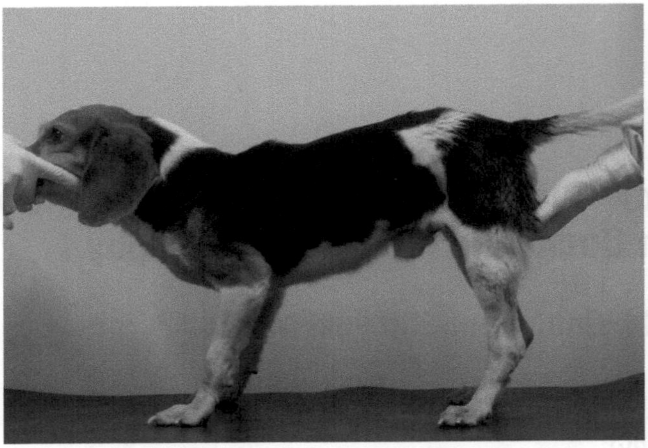

Fig. 1 A dog model with Duchenne muscular dystrophy (CXMDj)

Basically, these evaluations were performed under inhalation anesthesia and continuous pulse oximetry monitoring. All experiments are performed following the guidelines of NCNP Ethics Committee for The Treatment of Laboratory Middle-Sized Animals.

2 Materials

1. For the preparation of cardio-physiological examination, a 22-gauge or a 24-gauge Surflo needle with the drip infusion set including any isotonic solution is needed.
2. A 20 mg/kg of thiopental sodium is administered intravenously before endotracheal tube intubation.
3. For the usual size of beagle dog (CXMDj), from 4 kg to 10 kg of body weight, an endotracheal tube with 6.5 mm or 7.0 mm of internal diameter with a stylet and a Macintosh laryngoscope blade is used for intubation (Fig. 2a).
4. During the cardiac examination, inhalation anesthesia is maintained by 1.5–2.5% of isoflurane using the anesthesia apparatus (Fig. 2b).
5. The pulse oximetry system is used for vital sign monitoring such as pulse rate and oxygen saturation (Fig. 2c).
6. For cardio-physiological evaluations, each equipment is needed such as digital echocardiography system (Fig. 2d), 12-lead ECG recorder (Fig. 2e), 24-h Holter ECG recorder (Fig. 2f), and MRI system (Fig. 2g) including cardiac MRI software, MRI-compatible electrodes, and MRI-compatible artificial ventilator (Fig. 2h).

Fig. 2a Materials for endotracheal tube intubation

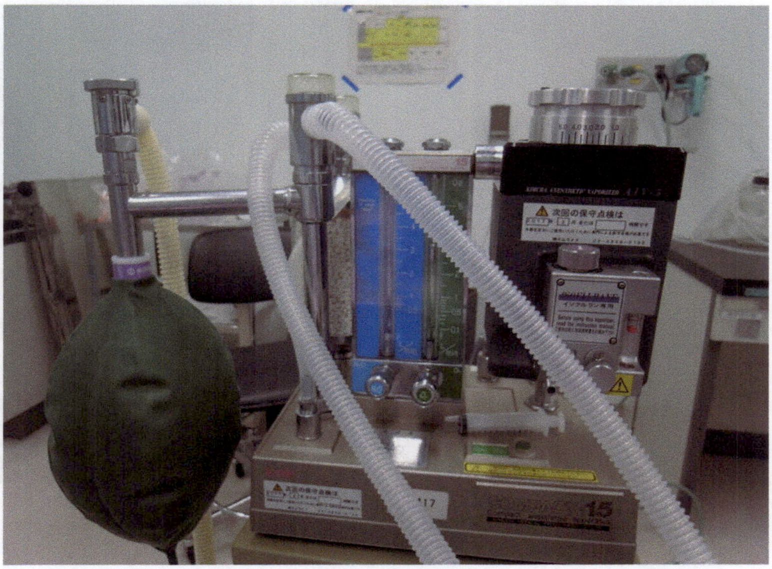

Fig. 2b Inhalation anesthesia apparatus

7. For the optional late gadolinium enhancement (LGE) MRI imaging, 2 mL/kg of gadolinium contrast media is necessary.
8. For echocardiography, we use a customized table with a rectangularly indented (arrow in Fig. 2i) shape to regulate the probe easily in the left lateral decubitus position.

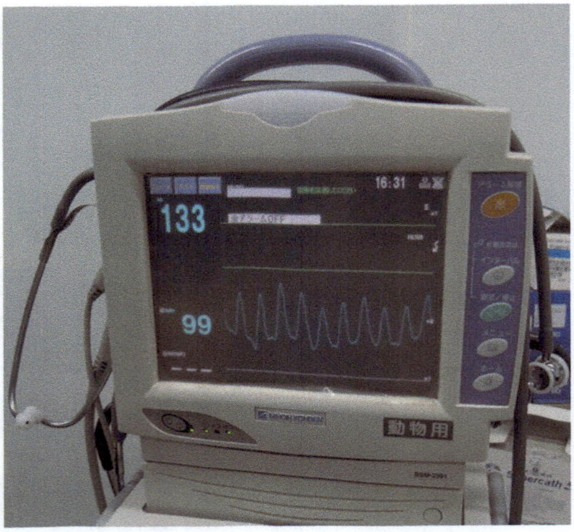

Fig. 2c Pulse oximetry system for animal experiments

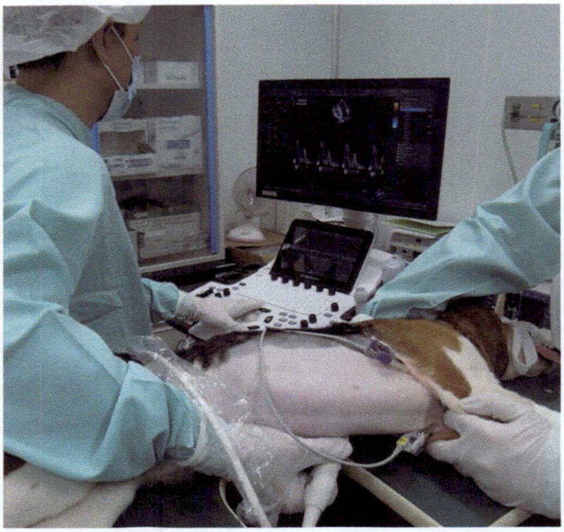

Fig. 2d Digital echocardiography system

3 Methods

3.1 Preparation

1. Dogs were starved after feeding in the evening on the day before the examination. Water can be provided until the morning of the examination date.
2. After keeping a vascular access route in the foreleg using a Surflo needle, thiopental sodium is administered intravenously.
3. For endotracheal tube intubation, the dog should be in prone position, and an assistant should open the mouth wide to help the tube intubator (Fig. 3a).

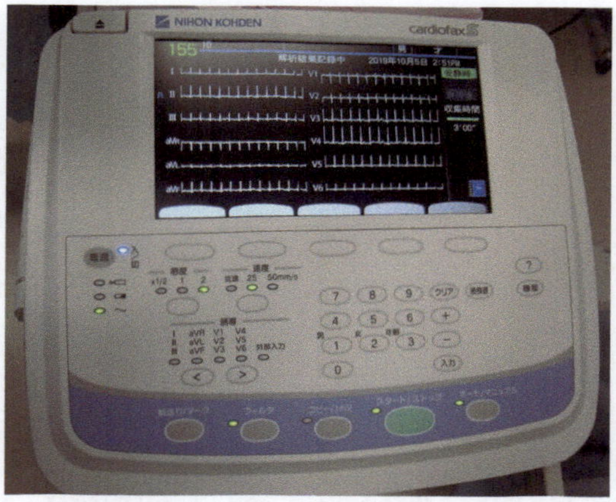

Fig. 2e Twelve-lead ECG recorder

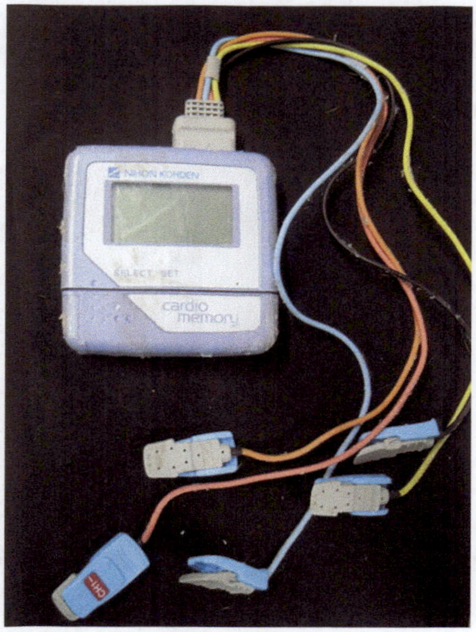

Fig. 2f Twenty-four hours Holter ECG recorder

4. After intubation (*see* **Note 1**), pay attention as the intubation tube will be fogged up with exhalation, and use a stethoscope to confirm evident bilateral breath sounds.
5. Place a pulse oximeter on the tongue to check the pulse and oxygenation (Fig. 3b).
6. Shave the wide area of the front chest cleanly by a hair clipper (Fig. 3c).

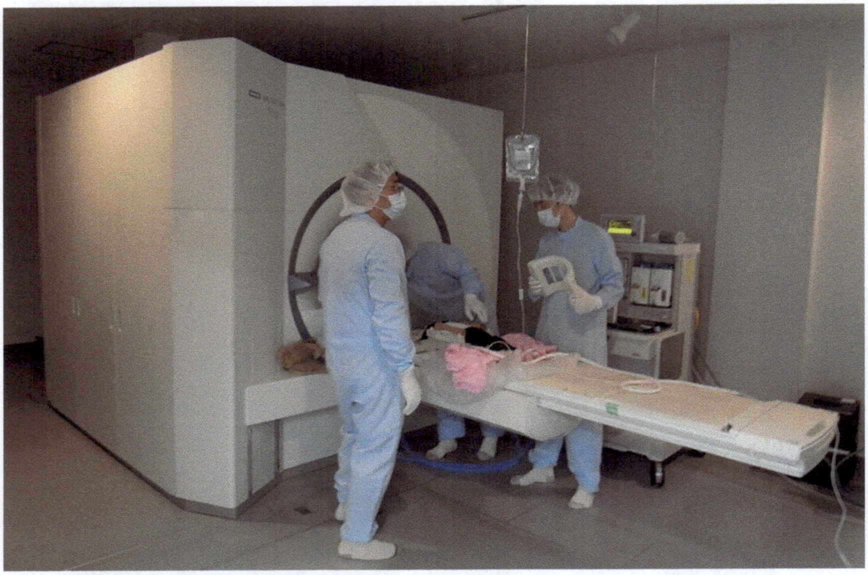

Fig. 2g MRI system for animal experiments

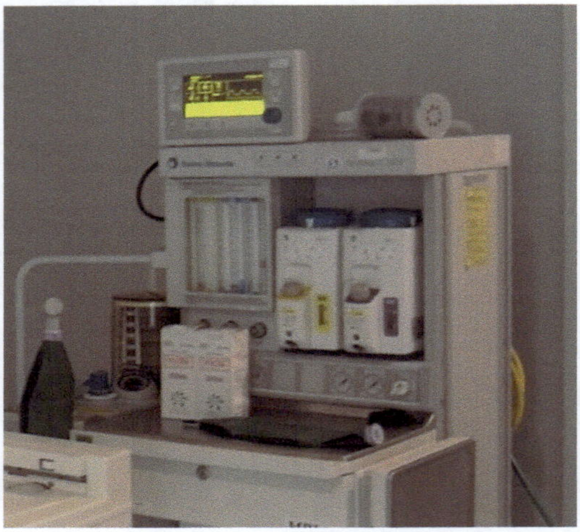

Fig. 2h MRI compatible artificial ventilator

3.2 Echocardiography

1. The animal should be kept in the left lateral decubitus position.
2. Spray 70% ethanol over a wide chest and wipe to degrease the skin surface (Fig. 3d).
3. Use a digital echocardiography system with a high-frequency pediatric sector probe for animal evaluation (Fig. 3e).
4. In the parasternal longitudinal view, measurements of aortic diametar (Ao) and left atrial diameter (LA) (Fig. 3f) can be skipped in the general animal experiments. However, before the step of left ventricular (LV) measurements, visual

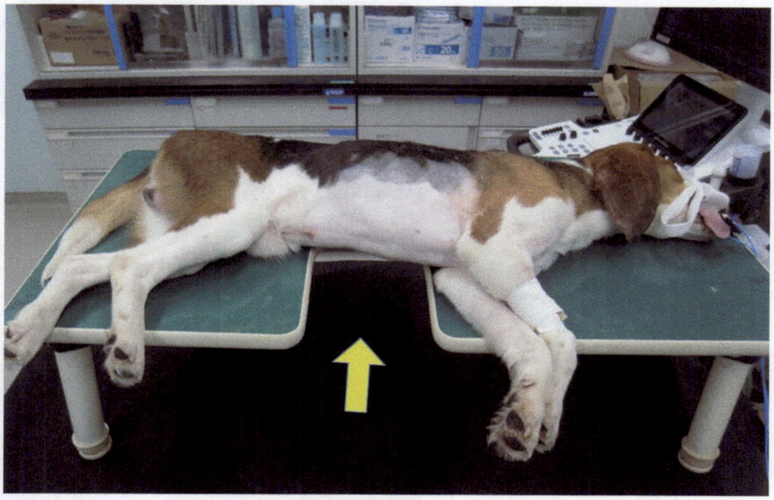

Fig. 2i A customized table for animal echocardiography

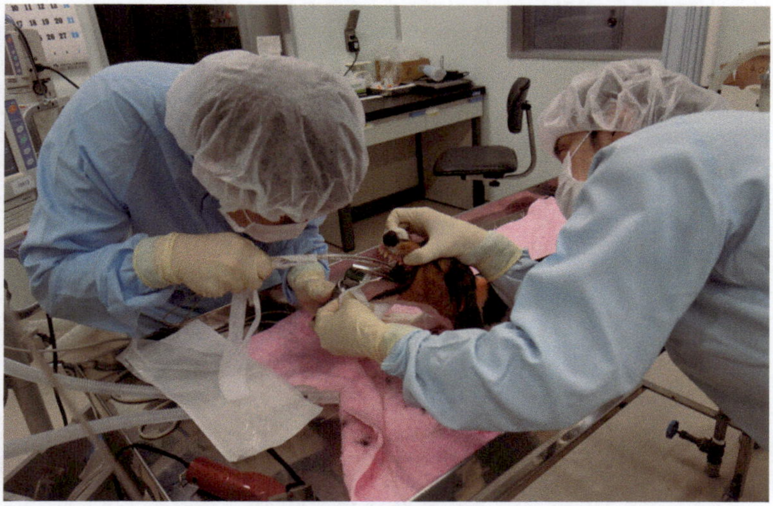

Fig. 3a Endotracheal tube intubation to the dog

evaluations of global LV contraction in the parasternal shot-axis view (Fig. 3g) is strongly recommended.

5. In the parasternal longitudinal view, measure LV diameters (*see* **Note 2**) in 2D image (Fig. 3h) as well as M-mode image (Fig. 3i) to obtain LV diameter at end-diastole (LVDd), LV diameter at end-systole (LVDs), LV fractional shortening (LVFS), and LV ejection fraction (LVEF) by Teichholz method.
6. In the apical 4-chamber view (Fig. 3j), check the presence of significant valvular diseases (*see* **Note 3**) using color Doppler

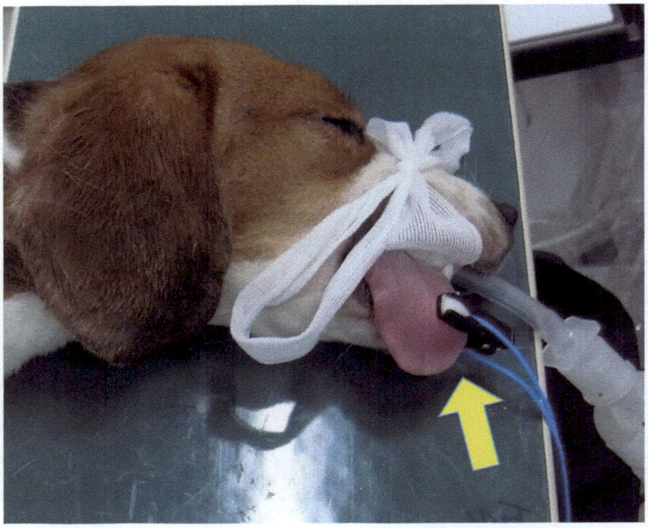

Fig. 3b Place a pulse oximeter on the tongue

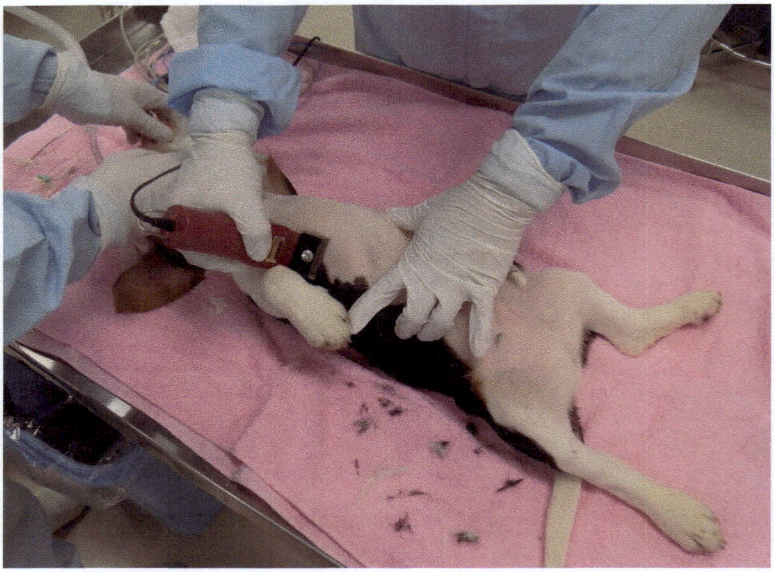

Fig. 3c Shave the front chest by a hair clipper

imaging. If necessary, the right ventricular (RV) functional measurements such as tricuspid annular plane systolic excursio-can (TAPSE) by M-mode image and RV fractional area change (RVFAC) by tracing RV area can be added.

7. When applicable, measurement of a single-plane Simpson's LVEF (Fig. 3k) by tracing the LV area in the apical 4-chamber view is recommended (*see* **Notes 4** and **5**).

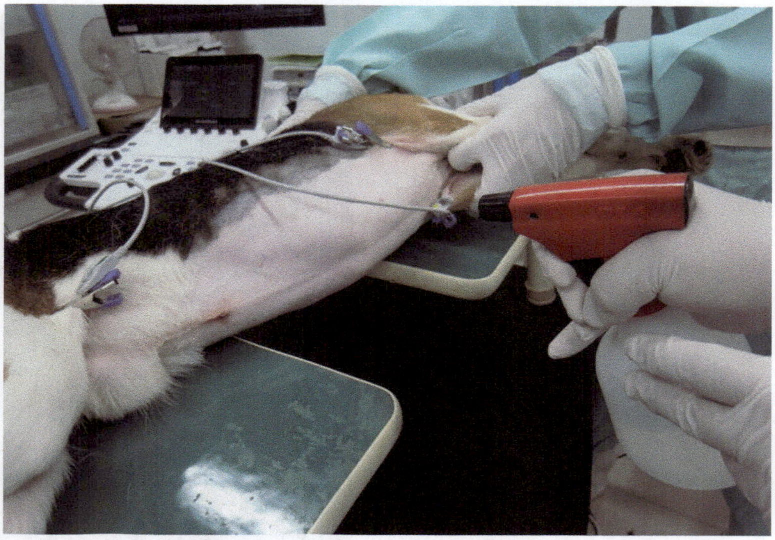

Fig. 3d Spray 70% ethanol to degrease the skin surface

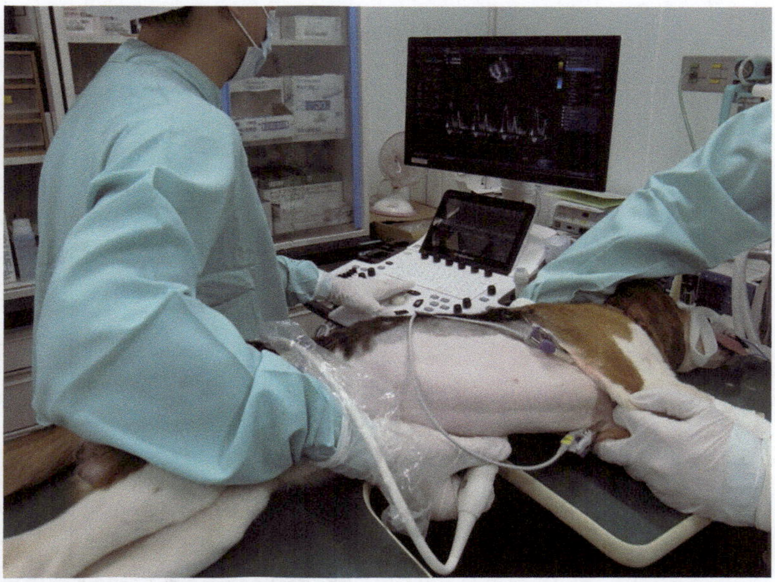

Fig. 3e Use a digital echocardiography system with a high frequency pediatric sector probe

8. In the apical 4-chamber view, evaluations of Doppler imaging (Fig. 3l) and tissue Doppler imaging (Fig. 3m) are recommended to evaluate LV diastolic functions.

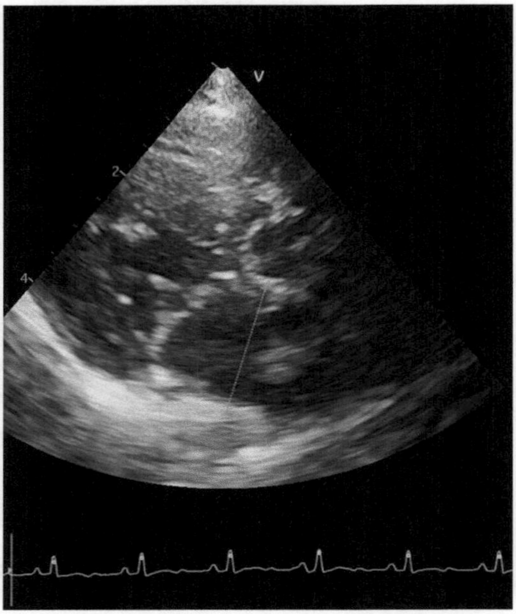

Fig. 3f Parasternal long-axis view of dog echocardiography

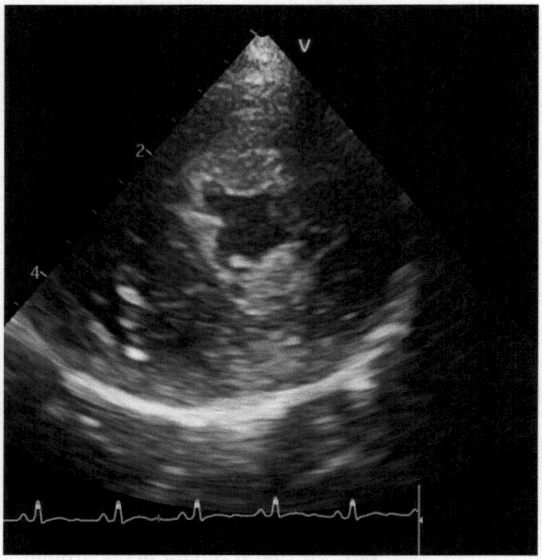

Fig. 3g Parasternal shot-axis view of dog echocardiography

3.3 Twelve-Lead ECG

1. Mark every chest electrode position with a permanent marker (Fig. 3n).
2. V1 electrode is placed at the right of the sternum at the fifth intercostal space.
3. The sixth intercostal space is used for all left-sided (V2–V6) leads.

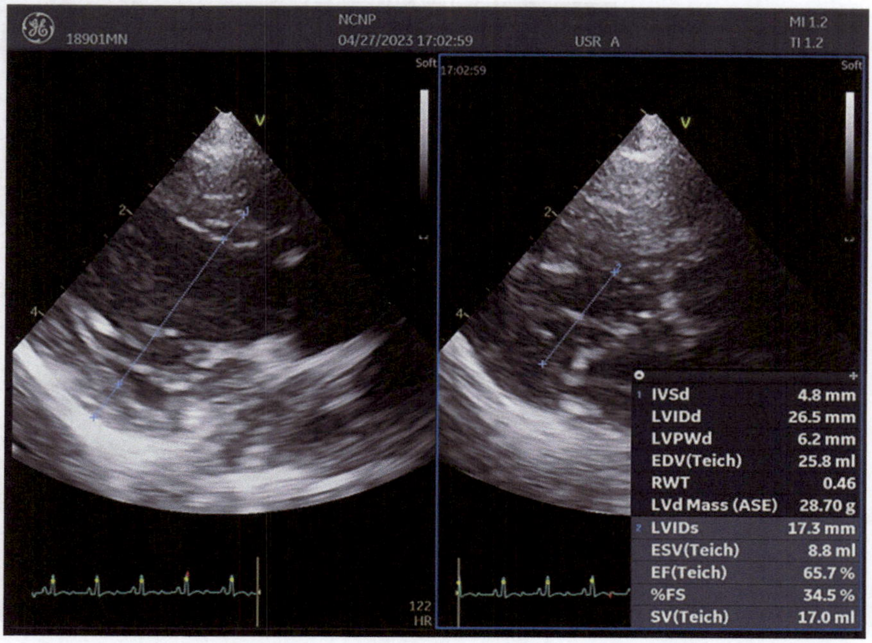

Fig. 3h Left ventricular measurements in parasternal 2D images

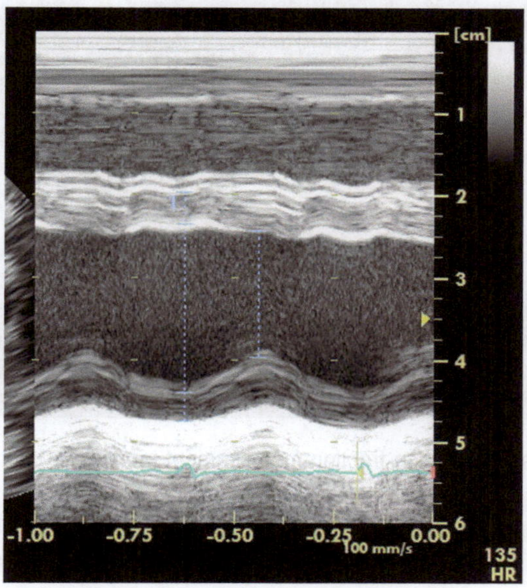

Fig. 3i Left ventricular measurement in M-mode image

4. V2 electrode is placed just at the left of the sternum.
5. V4 electrode is placed at the costochondral junction.
6. V3 electrode is placed at the middle position between V2 and V4.

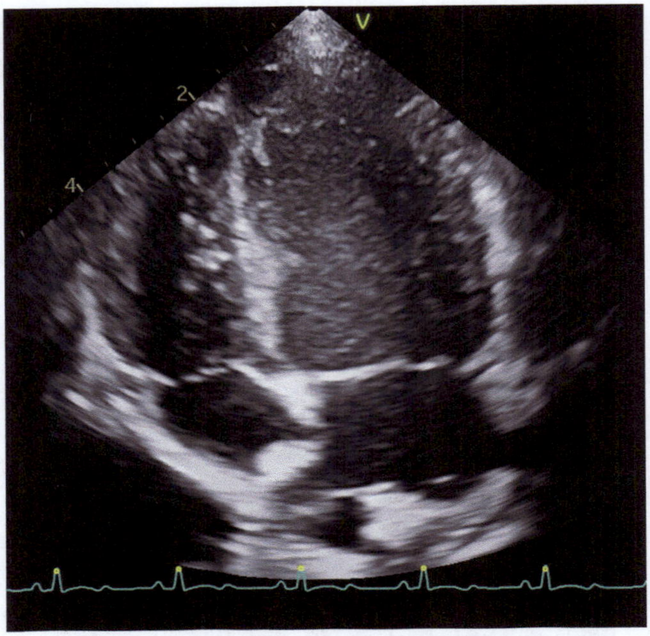

Fig. 3j Apical 4-chamber view of dog echocardiography

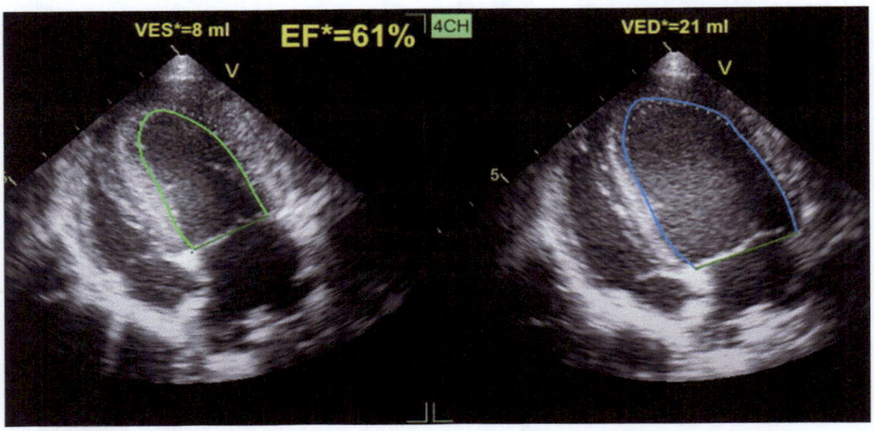

Fig. 3k Simpson's ejection fraction measurement in apical 4-chamber view

7. V6 electrode is placed at the intersecting point of left midaxillary line and sixth intercostal space.
8. V5 electrode is placed at the middle position between V4 and V6.
9. Limb lead electrodes are placed at each limb surrounding the chest (Fig. 3o).

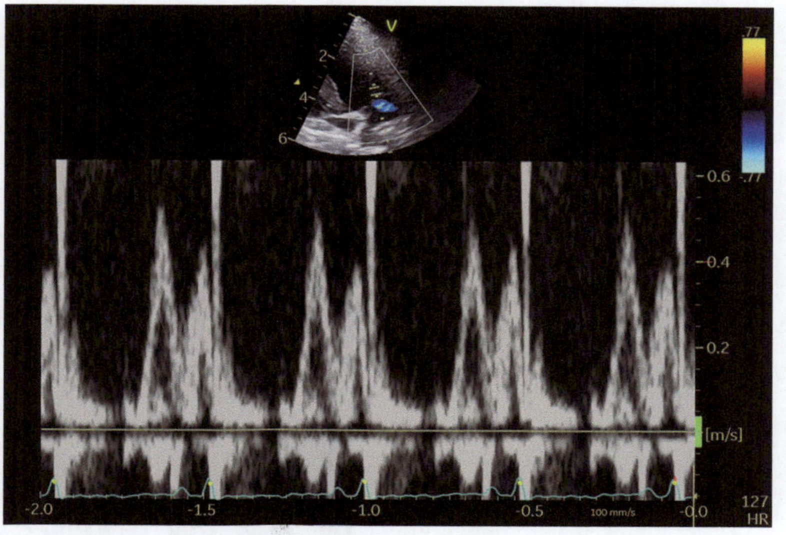

Fig. 3l Doppler imaging of dog echocardiography

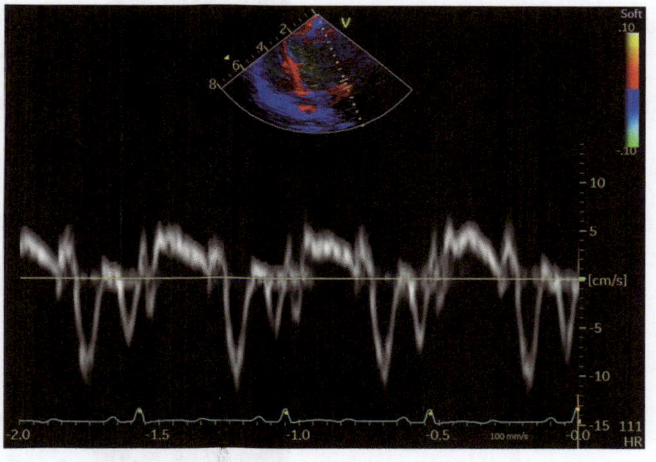

Fig. 3m Tissue Doppler imaging of dog echocardiography

10. Connect all electrodes to the electrocardiogram equipment and confirm all noises (arrow in Fig. 3p) are settled in every waveform.

11. Although the right lateral decubitus position (Fig. 3q) is recommended for 12-lead ECG recording in general veterinary text, we record electrocardiogram in supine position because we think that the positional relations between heart and chest electrodes are relatively fixed in that position. Hence, we can acquire high reproducibility results in repeated examinations (*see* **Note 6**).

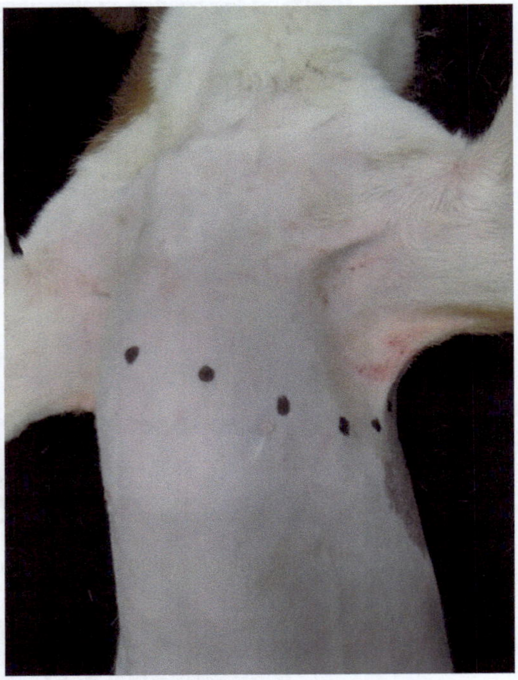

Fig. 3n Mark every electrode position in supine position

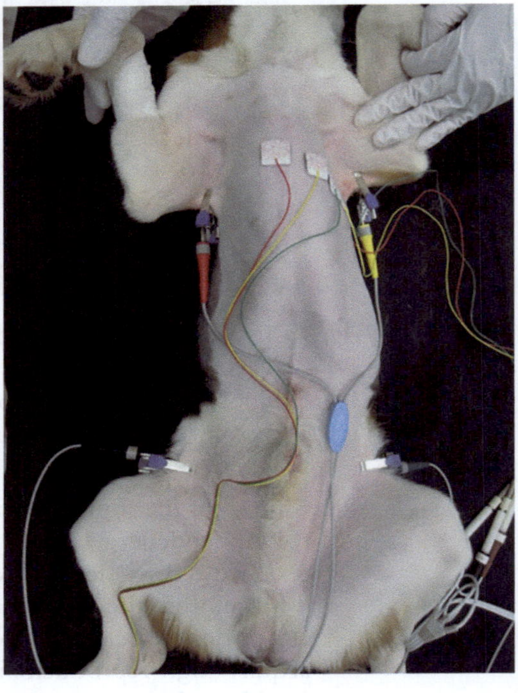

Fig. 3o Place limb lead electrodes on each limb

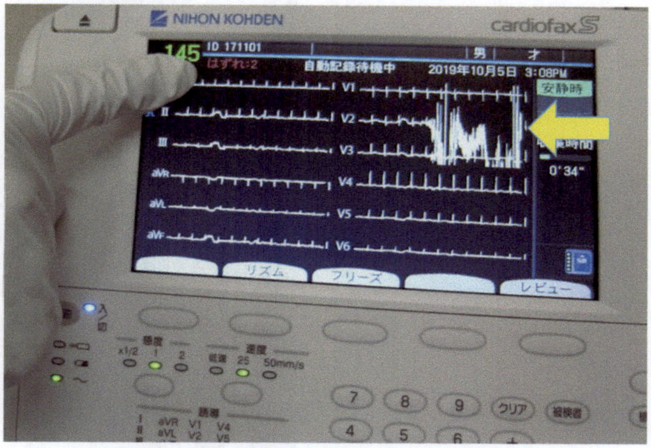

Fig. 3p Confirm no noise is found in all electrocardiogram waveforms

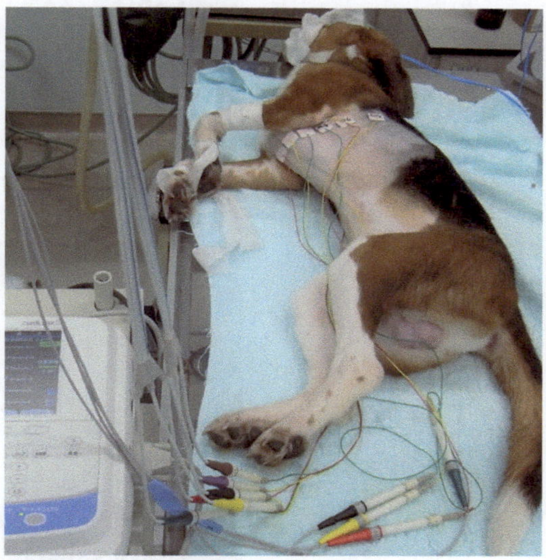

Fig. 3q Recoding electrocardiogram in right lateral decubitus position

3.4 Twenty-Four Hour Holter ECG

1. Place Holter ECG electrodes at the anterior chest surrounding the heart (Fig. 3r).
2. Wrap and fix the electrode leads with a tape bandage (Fig. 3s).
3. Wear a customized jacket for Holter ECG examination and put the device into the pocket (Fig. 3t).
4. Wear a neck cover to prevent from biting device (Fig. 3u).
5. Turn on the device to start recording ECG for 24 h.
6. The analyzed Holter ECG data provides informations regarding heart rate counts, arrythmia, and heart rate variabilities that reflect autonomic nervous system functions (*see* **Note** 7).

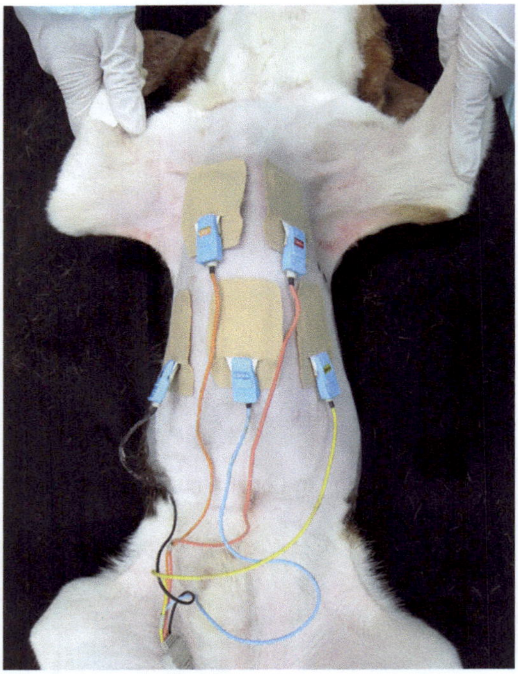

Fig. 3r Place Holter electrocardiogram electrodes

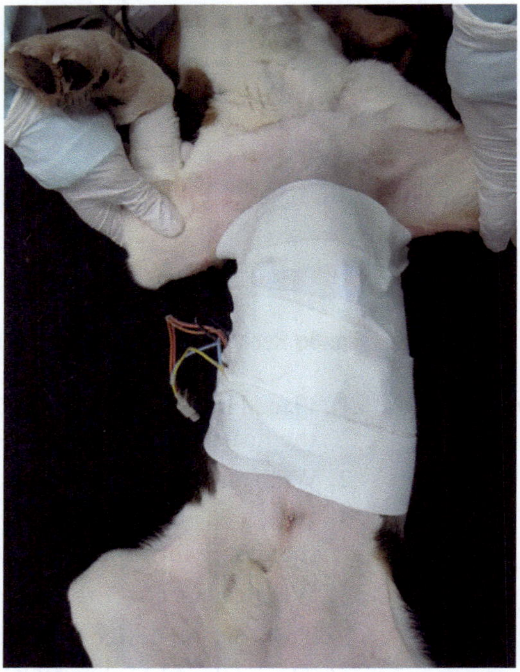

Fig. 3s Wrap and fix the electrode leads

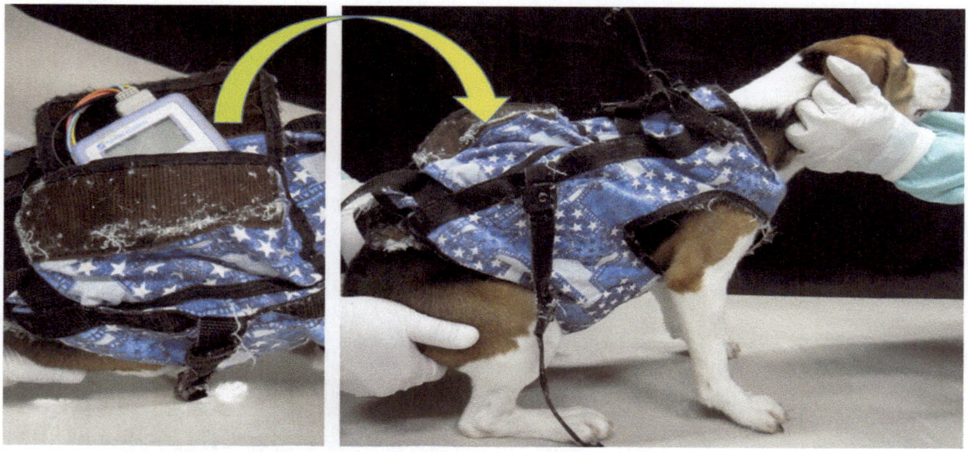

Fig. 3t Wear a customized jacket and put the Holter electrocardiogram device into the pocket

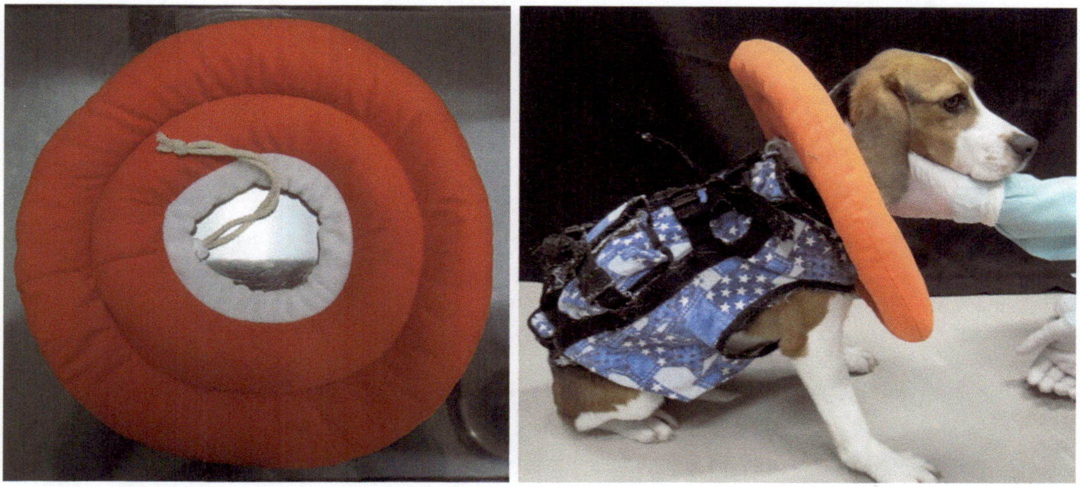

Fig. 3u Wear a neck cover to prevent from biting device

3.5 Cardiac MRI

1. Place MRI compatible electrodes at the chest (Fig. 3v).
2. Wrap and fix the electrode leads with tape bandages and connect to the MRI electrode cord (Fig. 3w).
3. Set a cardiac radiofrequency coil around the chest on the MRI bed (Fig. 3x).
4. The cardiac MRI can evaluate cardiac structure, dimension, volume, and function (Fig. 3y).
5. Optional evaluation of late gadolinium enhancement (LGE) imaging can be acquired (Fig. 3z) after 2 mL/kg of gadolinium contrast media injection for evaluating myocardial fibrosis (*see* **Note 8**).

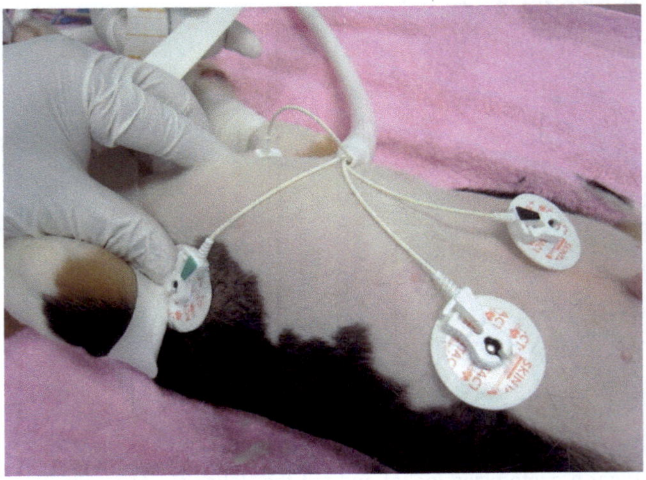

Fig. 3v Place 4 MRI electrodes

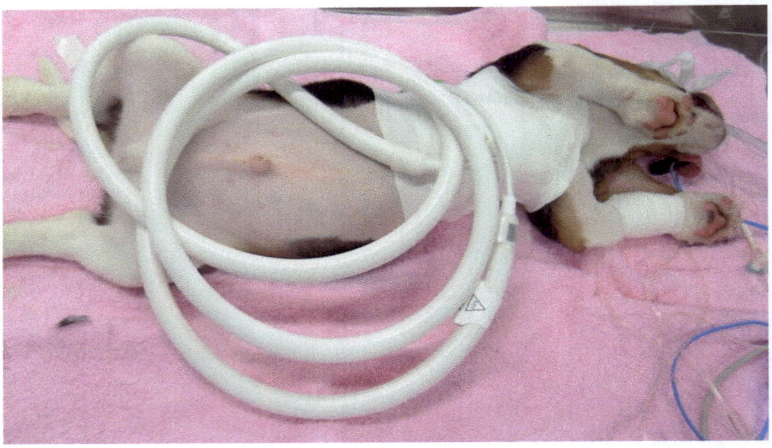

Fig. 3w Wrap the whole chest and connect to the MRI electrode code

4 Notes

1. Keep eyes on the depth of intubation tube into the trachea until the procedure is complete, when the assistant withdraws the stylet (Fig. 4a).
2. Note that the baseline results of LV diameter increase with age until 2 years of age even in the wild-type dogs (Fig. 4b). This result indicates that LV remodeling and dilatation can be considered after 2 years of age in this DMD model. Furthermore, in general experimental animals, LV diameter varies widely between individuals and can be affected by other factors such

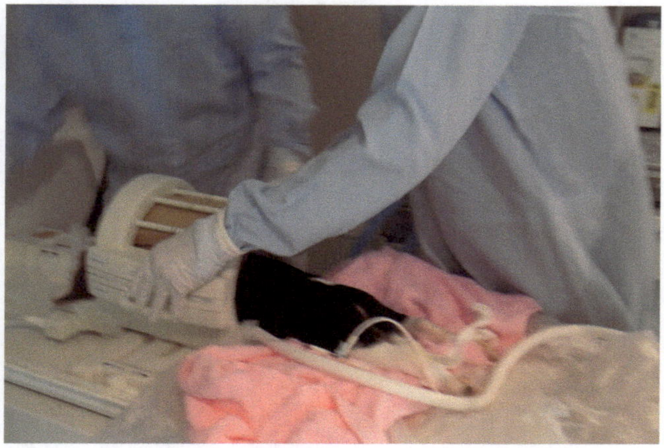

Fig. 3x Set a cardiac radiofrequency coil around the chest on the MRI bed

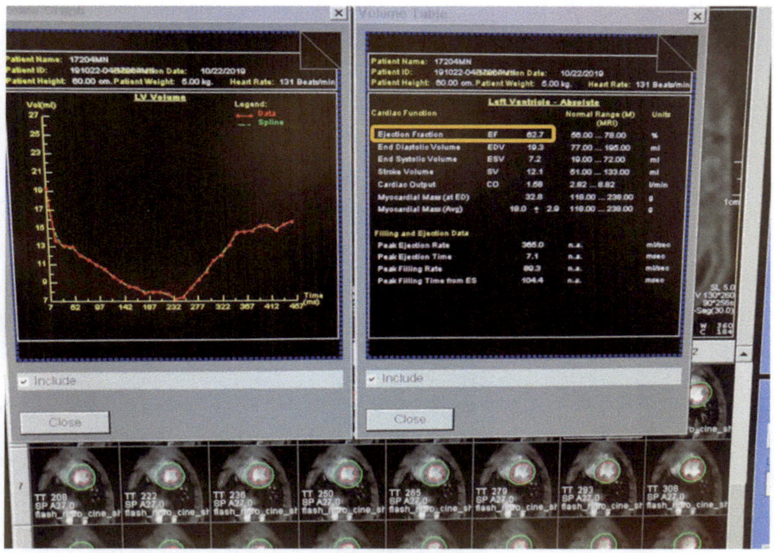

Fig. 3y Cine MRI imaging to evaluate cardiac function.

as feed intake, body size, and weight. We recommend to include a control group in the experimental protocol for the evaluation and interpretation of animal LV enlargement and remodeling.

3. This evaluation step can be skipped because a significant (more than mild) valvular disease is rarely present in this DMD dog model.
4. Although the individual left ventricular ejection fraction (LVEF) value varies widely, the slope angles of the approximation formula show 1.9% point per year decrease in DMD dogs (CXMDj), whereas 1.0% point per year decrease in female

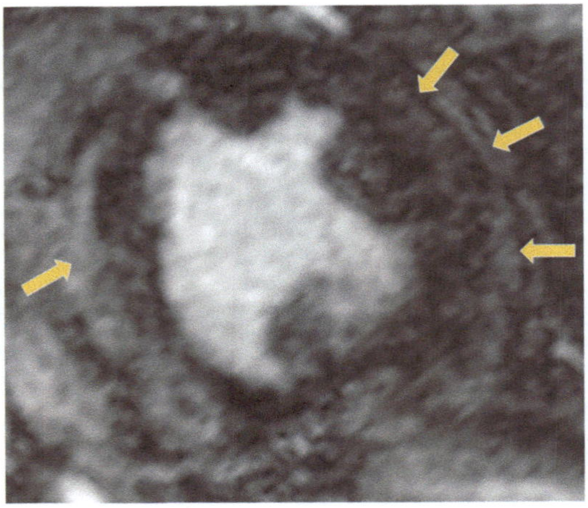

Fig. 3z Late gadolinium enhancement (LGE) imaging. In the MRI short-axis view, white line and spot lesions by LGE imaging indicate fibrotic degeneration (yellow arrows)

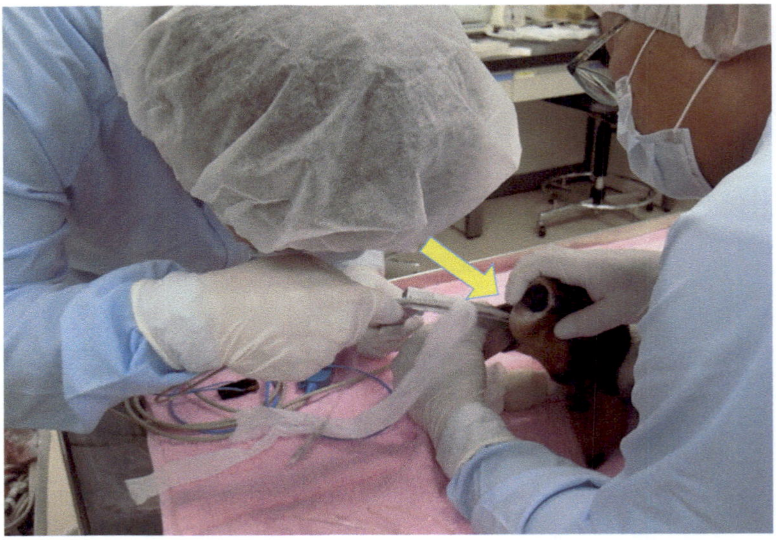

Fig. 4a Keep eyes on the depth of intubation tube until the procedure is complete

heterozygote carrier dogs, whose progression speed is half as fast as those in DMD dogs (Fig. 4c).

5. In quadruped animals, the apical 4-chamber view can be obliquely imaged, which often cause difficulty to trace left ventricular (LV) area to measure LVEF by Simpson's method. Because, the apical 2-chamber view is generally more inappropriately imaged, we recommend using 4-chamber view single-plane Simpson's method to measure LVEF.

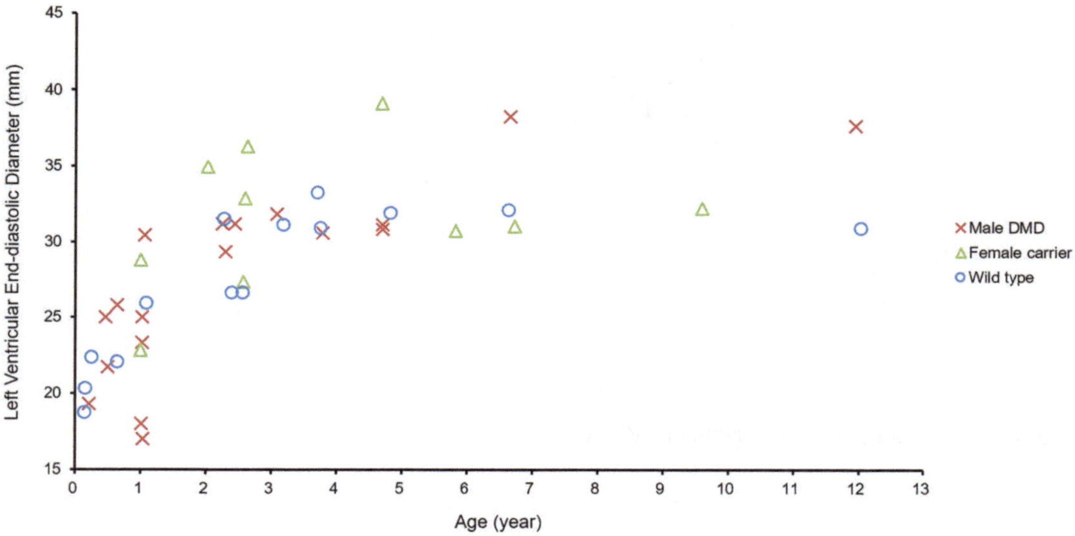

Fig. 4b Age-dependent change of left ventricular end-diastolic diameter evaluated by echocardiograpy. Blue circles indicate wild-type dogs ($n = 14$), green triangles indicate female heterozygote carrier dogs ($n = 10$), and red crosses indicate male DMD dogs ($n = 18$)

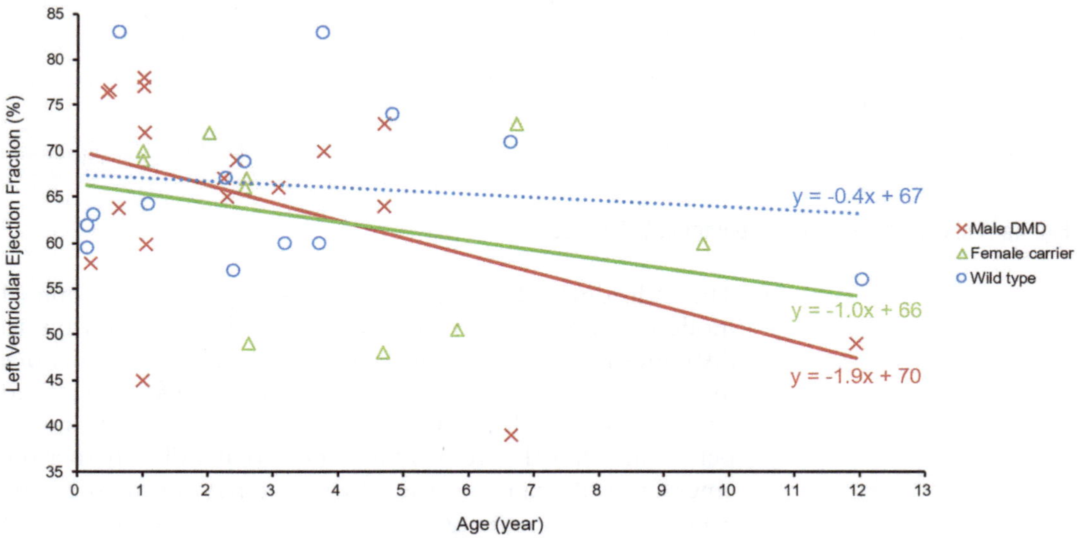

Fig. 4c Age-dependent change of left ventricular ejection fraction (LVEF) evaluated by echocardiography. Color lines indicate linear approximate formula for each group. Blue line and circles indicate wild-type dogs ($n = 14$), green line and triangles indicate female carrier dogs ($n = 10$), and red line and crosses indicate male DMD dogs ($n = 18$). Although the individual LVEF value varies widely, the slope angles of the approximation formula show 1.9% point per year decrease in DMD dogs (CXMDj), whereas 1.0% point per year decrease in female heterozygote carrier dogs, whose progression speed is half as fast as those in DMD dogs

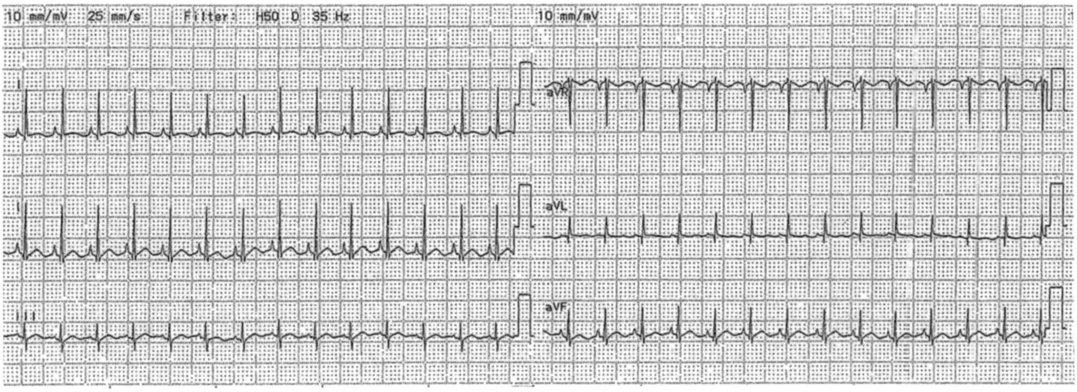

Fig. 4d A limb lead electrocardiogram in DMD dogs

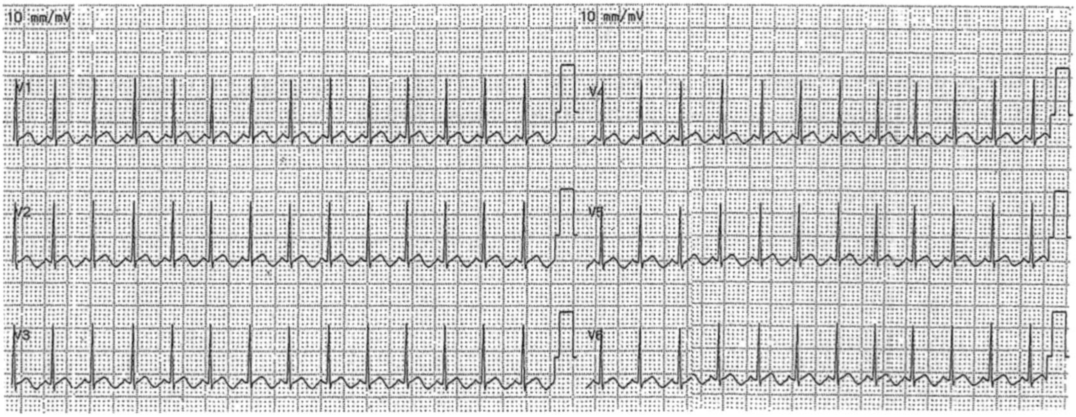

Fig. 4e A chest lead electrocardiogram in DMD dogs

6. The individual 12-lead ECG findings (Fig. 4d, e) in animal models vary widely. In most cases, it might be difficult to distinguish abnormal findings from normal-range variations in a single evaluation of 12-lead ECG. Therefore, repeated evaluations of 12-lead ECG are recommended at the timings before and after the study intervention to identify whether the emergence of change in their ECG findings is caused by the experimental intervention. For the clear detection of this change, we recommend the recording of ECG in supine position and mark the all-electrode positions with a permanent marker to keep robust reproducibility.
7. The mean heart rate and minimum heart rate by 24 hour Holter ECG are higher in female heterozygote carrier dog and male DMD dog (CXMDj) than those in wild-type dogs (Fig. 4f). The heart rate variability analysis reveals that SDANN (standard deviation of the normal–normal R–R period) and HF (high-frequency spectra) are decreased in female

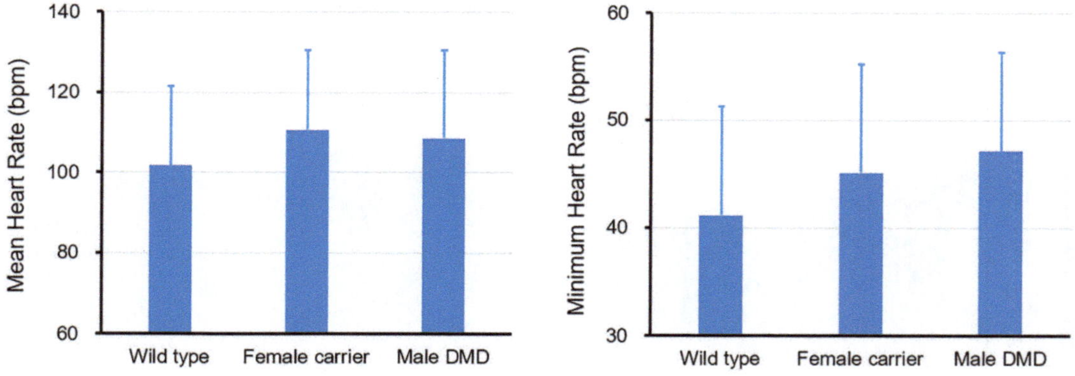

Fig. 4f Heart rate results from a 24-hour Holter electrocardiogram in dogs. The bar graphs indicate mean ± SD values. Results are from age-matched 24-hour Holter ECG tests from wild-type dogs (1.8 ± 2.2 years of age, $n = 35$), female heterozygote carrier dogs (1.8 ± 2.1 years of age, n = 25), and male DMD dogs (1.8 ± 2.0 years of age, $n = 43$)

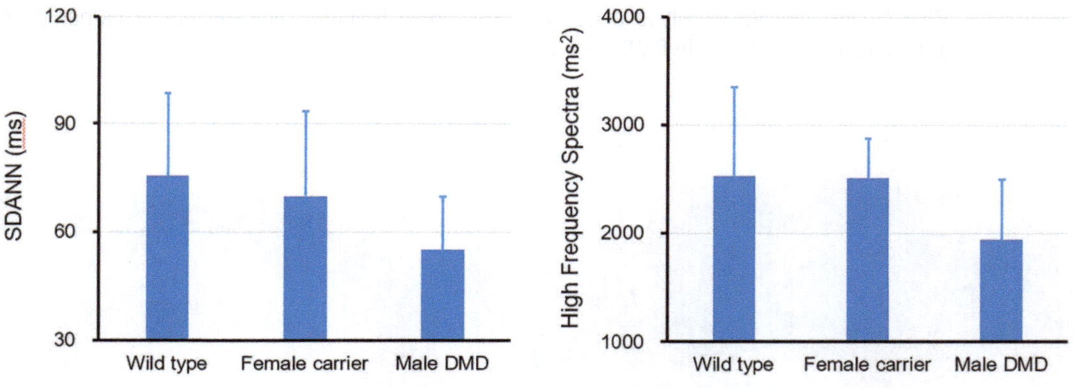

Fig. 4g Parasympathetic activity from heart rate variability analysis in dogs. The bar graphs indicate mean ± SD values. The heart rate RR variability analysis results from age-matched Holter ECG tests from wild-type dogs (1.8 ± 2.2 years of age, $n = 35$), female carrier dogs (1.8 ± 2.1 years of age, $n = 25$), and male DMD dogs (1.8 ± 2.0 years of age, $n = 43$). SDANN = standard deviation of the average N-N intervals

heterozygote carrier dog and male DMD dog (Fig. 4g), the variables of which are related to parasympathetic activity. It is noted that increased heart rate is observed even in young wild-type dogs before ages of 1.5 years (Fig. 4h).

8. Fibrotic degenerations are detected as white lesions in late gadolinium enhancement (LGE) imaging (arrows in Fig. 3z), whose findings are consistent with those from the Masson-trichrome staining of the same dog's necropsied heart (Fig. 4i).

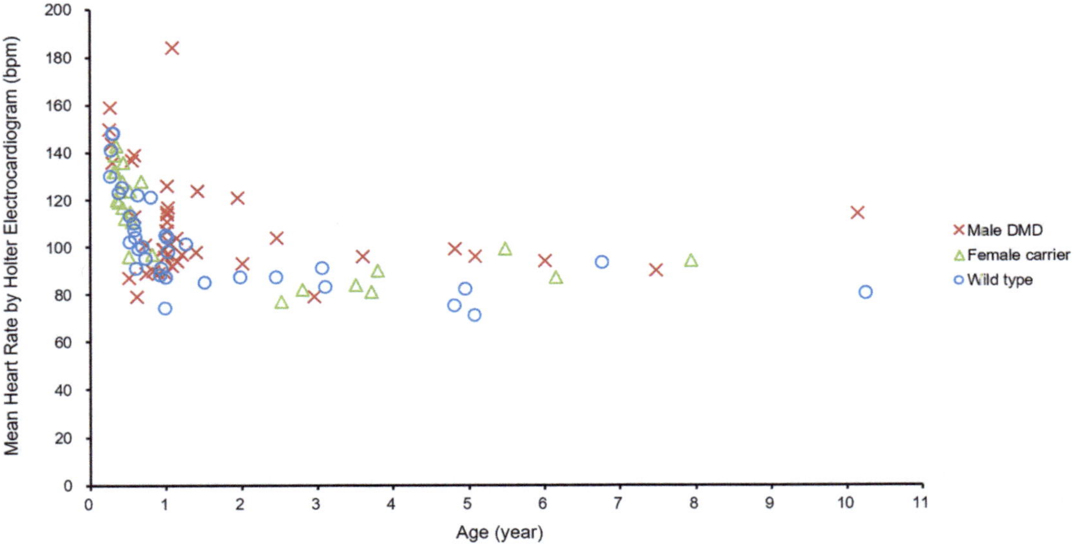

Fig. 4h Age-dependent change of mean heart rate in dogs. The mean heart rate was evaluated by 24-hour Holter ECG. Blue circles indicate wild-type dogs ($n = 35$), green triangles indicate female carrier dogs ($n = 25$), and red crosses indicate male DMD dogs ($n = 43$)

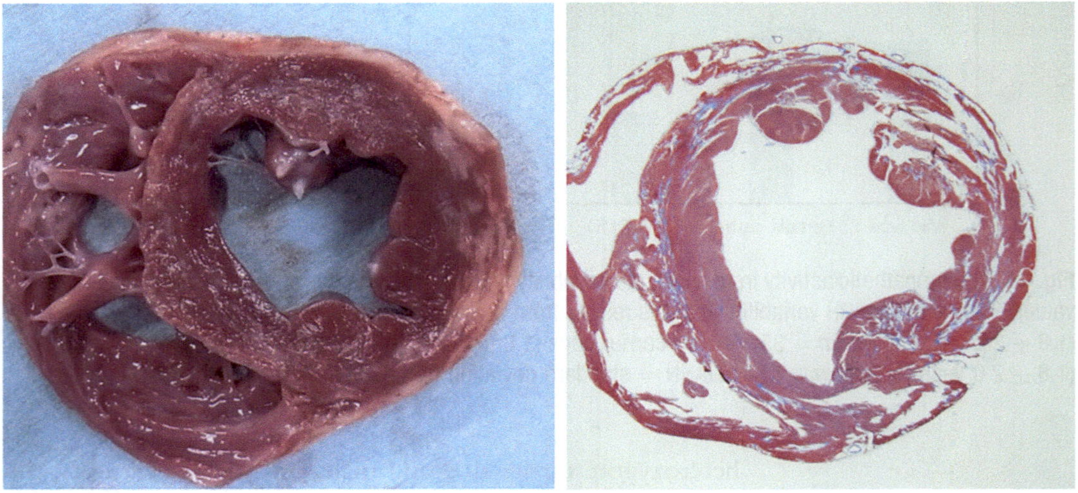

Fig. 4i Heart necropsy and pathological study of the same DMD dog that underwent MRI (LGE imaging in Fig. 3z). Representative images of heart necropsy (left panel) and pathological study by Masson-trichrome staining (right panel). This DMD dog was necropsied at 5 years old (61 months old) and also evaluated by MRI (Fig. 3z) at 3 months before the necropsy

Acknowledgments

This work was supported by JPJS KAKENHI 23 K06905 and AMED JP21ek0109496.

Conflict of Interest The authors have no conflicts of interest to declare that are relevant to the content of this chapter.

Chapter 4

Splice Modulation Studies Using Urine-Derived Cells

Katsuhiko Kunitake and Yoshitsugu Aoki

Abstract

Antisense oligonucleotide (ASO)-based exon skipping is a representative splice modulating therapy effective for Duchenne muscular dystrophy (DMD), a condition arising from dystrophin deficiency. For a valuable platform of evaluating exon-skipping efficacy, we previously reported a highly effective approach to convert CD90-positive human urine-derived cells (UDCs) into myotubes through the transduction of the *MYOD1* gene. We anticipate that highly differentiated myotubes enable their utilization in various human muscle cell studies and a more predictive screening approach for the preclinical identification of promising ASOs at an earlier stage of the drug discovery process. Here, we describe protocols and tips for isolating UDCs, differentiating CD90-positive UDCs into myotubes, and evaluating the correction of *DMD* mRNA and protein levels in the myotubes after exon skipping.

Key words Duchenne muscular dystrophy (DMD), Human urine-derived cells (UDCs), CD90, Antisense oligonucleotide (ASO), Exon skipping, MYOD1, RT-PCR, Immunoblotting, Immunocytochemistry

1 Introduction

Duchenne muscular dystrophy (DMD) is a progressive muscular disease with severe muscle weakness caused by frame-shift mutations in the *DMD* gene leading to the lack of dystrophin protein [1]. One of the promising therapies for DMD is the antisense oligonucleotide (ASO)-based exon skipping. This exon skipping has the capability of modulating *DMD* pre-mRNA to shift an out-of-frame deletion in-frame and restore the truncated but functional dystrophin protein [2]. In fact, we have reported that viltolarsen, an exon 53-skipping ASO drug, possesses a promising efficacy, excellent safety profile, and acceptable pharmacokinetic parameters by weekly intravenous administration in a first-in-human study (registered as UMIN: 000010964 and ClinicalTrials.gov: NCT02081625) [3]. However, in this clinical trial, we realized that the effectiveness of ASOs was variable depending on each patient. Therefore, acquiring patient-derived primary cultured

Yoshitsugu Aoki and Norio Motohashi (eds.), *Splice Switching: Methods and Protocols*, Methods in Molecular Biology, vol. 2963,
https://doi.org/10.1007/978-1-0716-4738-7_4,

cells is crucial for validating individual response to the drugs before administration. In this context, we have focused on human urine-derived cells (UDCs) collected noninvasively and repeatedly from DMD patients [4, 5]. Notably, by using CD90-positive UDCs and transducing *MYOD1* gene to them, we have developed a high-quality primary myotube model with high and stable dystrophin expression levels [6]. Patient-derived UDC myotubes may thus contribute to a predictive screening strategy for effective new ASOs. Here, we describe a detailed protocol for isolating UDCs from voided urine, generating myotubes via *MYOD1* induction of CD90-positive UDCs, and evaluating exon-skipping efficiency and dystrophin restoration level in ASO-treated UDC myotubes from DMD patients.

2 Materials

2.1 Isolation and Culture of UDCs

1. Gelatin-coated MICROPLATE 6 well with a lid (IWAKI) for culturing UDCs.
2. Sterile plastic bottles for collecting urine.
3. Sterile PBS without calcium and magnesium.
4. Growth medium (GM) used for culturing UDCs: 1:1 mixture of REGM bullet kit (Lonza) and high-glucose Dulbecco's modified Eagle's medium (DMEM) (GE Healthcare) supplemented with 15% tetracycline-free fetal bovine serum (FBS; Clontech), 0.5% Glutamax (Thermo Fisher Scientific), 0.5% nonessential amino acids (Thermo Fisher Scientific), 2.5 ng/mL basic fibroblast growth factor (bFGF; Sigma-Aldrich), 2.5 ng/mL human PDGF-AB (Peprotech), 1% penicillin/streptomycin (P/S), and 0.5 μg/mL amphotericin B.
5. 0.25% Trypsin-EDTA (Gibco).
6. Gelatin-coated 60 mm dish (IWAKI) for culturing the UDCs after passage.
7. CELLBANKER® 1 (ZENOGEN PHARMA) for freezing the urine sample.

2.2 Sorting CD90-Positive UDCs

1. Sterile PBS without calcium and magnesium.
2. A 5 mL polystyrene round-bottom tube with a 40 μm cell-strainer cap (Corning) for analyzing and sorting.
3. Fluorescein isothiocyanate (FITC)-conjugated antihuman CD90 antibody (1:200, BioLegend, #328107).
4. A 5 mL polystyrene round-bottom tube (Corning) for collecting sorted UDCs.
5. GM described in Subheading 2.1.

6. A 100 μm sorting chip (Sony).
7. SONY FACS SH800S (Sony).
8. Cell Sorter Software (Sony).

2.3 Retroviral Construct

1. *MYOD1* (NM_002478) human tagged ORF clone (ORIGENE).
2. CloneAmp HiFi PCR premix (Clontech).
3. Forward and reverse primers targeted, respectively, on both sides across the inserted *MYOD1* sequence.
4. Sterilized distilled water.
5. 0.7% agarose gel.
6. NucleoSpin gel and PCR clean-up kit (Clontech).
7. NanoDrop (Thermo Fisher) as a spectrophotometer.
8. 500 ng/μL pRetroX-TetOne-Puro Vector (Clontech) containing a Tet-on system and *puromycin-resistant gene*.
9. 10x K buffer (TAKARA) to digest the retroviral vector described in Subheading 2.3, **item 8**.
10. 2-15 U restriction enzymes: BamH1 and EcoR1.
11. In-fusion HD enzyme premix (Clontech).
12. *Escherichia coli* DH5α competent cells (TAKARA).
13. An LB culture plate consisting of 10 g/L bacto tryptone, 5 g/L bacto yeast extract, 5 g/L NaCl, 15 g/L bacto agar, and 50 mg/L ampicillin.
14. An LB culture medium without bacto agar.
15. HiSpeed plasmid purification kit (Qiagen).
16. GP2–293 packaging cells (Clontech).
17. 100 mm/collagen-coated dish (IWAKI) for culturing packaging cells stated above.
18. DMEM with 10% FBS.
19. pVZV-G capsid vectors.
20. Xfect transfection reagent (Clontech) containing cell-penetrating peptide.
21. Retro-X concentrator (Clontech).
22. MILTEX HV 0.45 μm filter (MERCK) for filtering the retrovirus supernatant.
23. One-Step TB Green PrimeScript RT-PCR kit (TAKARA) for titration of the retroviral vector.
24. Retrovirus titer set (TAKARA) for titration of the retroviral vector.

2.4 Infection with MYOD1-Retroviral Vector in CD90-Positive UDCs

1. A gelatin-coated 100 mm dish (IWAKI) for culturing CD90-positive UDCs.
2. Hexadimethrine bromide.
3. 1 μg/mL puromycin.
4. GM described in Subheading 2.1.

2.5 Myogenic Differentiation of MYOD1-Transduced CD90-Positive UDCs

1. Differentiation medium (DM): High-glucose DMEM with 1% L-alanine–L-glutamine, 5% horse serum, and 1% ITS liquid media supplement (Sigma-Aldrich).
2. 1 μg/mL doxycycline.
3. Collagen-coated microplate 12-well/flat bottom (IWAKI) for differentiated UDCs used for RNA or protein extraction.
4. Collagen-coated microplate 96-well/flat bottom (IWAKI) for differentiated UDCs used for immunocytochemistry.

2.6 Transfection of ASO into MYOD1-Transduced CD90-Positive UDCs

1. 1 mM ASO (25-mer, phosphorodiamidate morpholino oligomer, Gene Tools).
2. 1 mM Endo-Porter transfection reagent (Gene Tools).
3. 1 mM control antisense (Gene Tools).

2.7 Evaluation of Exon-Skipping Efficiency by RT-PCR

1. RNase-free water.
2. QIAGEN OneStep RT-PCR kit (Qiagen).
3. 5 mM primer pairs designed corresponding to the target exon and each patient's mutation.
4. RNeasy kit (Qiagen) for RNA extraction.
5. NanoDrop (Thermo Fisher) as a spectrophotometer.

2.8 Detection of Dystrophin by Immunoblotting

1. Radioimmunoprecipitation assay (RIPA) buffer (Thermo Fisher Scientific) containing cOmplete Protease Inhibitor Cocktail (Roche).
2. Pierce BCA protein assay kit (Thermo Fisher Scientific) for measuring the concentration of proteins.
3. NuPAGE LDS sample buffer (4x) (Invitrogen).
4. NuPAGE sample reducing agent (10x) (Invitrogen).
5. Deionized water.
6. NuPAGE Tris–acetate SDS running buffer (Invitrogen) containing 8.95 g/L tricine, 6.06 g/L Tris-base, and 1.0 g/L sodium dodecyl sulfate (SDS).
7. NuPAGE 3–8% Tris–acetate protein gels (Invitrogen).
8. XCell SureLock Mini-Cell (Invitrogen) for electrophoresis.
9. EzFastBlot HMW (Atto) as blotting buffer during semidrying.
10. Immobilon-P transfer membrane (PVDF) (MERCK).

11. Extra thick blot filter paper (Bio-Rad).
12. Methanol.
13. Semidry transfer apparatus.
14. Mouse anti-dystrophin (1:20, Leica; NCL-DYS1) as a primary antibody.
15. Mouse anti-α-tubulin (1:1200, Sigma-Aldrich; T6199) as a primary antibody.
16. Anti-mouse horseradish peroxidase-conjugated secondary antibody (1:50000, Cell Signaling Technology) as a secondary antibody.
17. PBS containing 0.1% Tween 20 (PBS-T) for washing PVDF or diluting antibodies.
18. 5% ECL prime blocking agent (Cytiva) in PBS-T.
19. ECL prime western blotting detection reagent (GE Healthcare) as a chemiluminescent detection reagent.
20. ChemiDoc MP Imaging System (Bio-Rad) as a charged-coupled device (CCD) camera-based imager.

2.9 Detection of Dystrophin by Immunocytochemistry

1. Sterile PBS without calcium and magnesium.
2. 4% paraformaldehyde.
3. 0.1% Triton-X (MP Biomedicals) as a nonionic detergent.
4. 10% goat serum diluted by PBS (10% GS) for blocking.
5. Mouse anti-dystrophin (1:20, Leica, NCL-DYS1) as a primary antibody.
6. Alexa Fluor 546 goat anti-mouse IgG (1:300, H + L; Invitrogen) as a secondary antibody.
7. DAPI solution (1:20000, DOJINDO) for nuclear staining.
8. BZ810 (KEYENCE) as a fluorescence microscope.
9. BZ-810 analyzer (KEYENCE) for analyzing the fluorescence signal intensity.

3 Methods

3.1 Isolation and Culture of UDCs (See Fig. 1)

1. Collect over 25 mL of midstream urine in a sterilized plastic bottle on day 0 (*see* **Note 1**).
2. Divide the entire volume of urine into the required number of 50 mL tubes.
3. Centrifuge tubes at 400× *g* for 10 min at room temperature.
4. Aspirate the supernatant, leaving 3 mL in each tube (*see* **Note 2**).

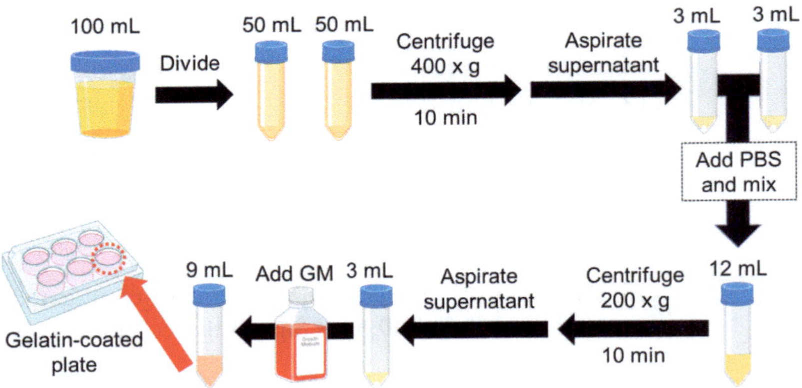

Fig. 1 Protocol of human urine-derived cell (UDC) isolation

5. Resuspend the pellet (*see* **Note 3**) in the remaining 3 mL of urine for each tube and then collect the cell suspension from different tubes into a single 50 mL tube.
6. Add an equal amount of PBS to the cell suspension and centrifuge it at 200× *g* for 10 min at room temperature.
7. Aspirate the supernatant, leaving 3 mL in the tube. If you need to stop the experiment here, add 3 mL of CELLBANKER® 1 in the tube. This total 6 mL liquid can be frozen and stored at −80 °C.
8. Resuspend the cell pellets in the remaining 3 mL of liquid mentioned above and add 6 mL of GM.
9. Seed the total cells in a single well of a gelatin-coated six-well plate (total volume of the well, 9 mL). Culture incubated at 37 °C and 5% CO_2 for 72 h.
10. On day 3, replace the medium with 2 mL of GM.
11. Change GM every other day (*see* **Note 4**).
12. When the UDC culture becomes 80–90% confluent, remove the medium and wash the cells with PBS, split all the cells using 0.25% trypsin–EDTA, and seed them onto a new gelatin-coated 60 mm dish.

3.2 Sorting CD90-Positive UDCs (See Fig. 2)

1. Trypsinize and wash UDCs with GM.
2. Centrifuge the UDC suspension at 200× *g* for 5 min.
3. Aspirate the supernatant and resuspend in PBS.
4. Add FITC-conjugated antihuman CD90 antibody to the suspension (1:200).
5. Incubate the suspension for 30 min on ice in the dark.
6. Centrifuge the suspension at 200× *g* for 5 min again.

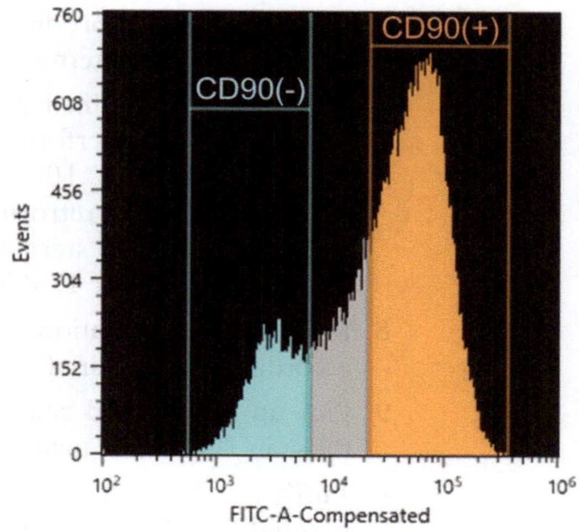

Fig. 2 Flow cytometric evaluation of UDCs using fluorescein isothiocyanate-conjugated antihuman CD90 antibody

7. Resuspend the pellet in PBS, pass through 40 μm filtering cap, and collect to the tube which is set for the cell sorter.
8. Analyze the proportion of CD90-positive UDCs in the sample.
9. Sort CD90-positive UDCs to the collection tube if the sample contains a large number of CD90-positive UDCs. If CD90-negative UDCs exist dominantly in the sample, it should be better to return to Subheading 3.1 (*see* **Note 5**).
10. Seed CD90-positive UDCs onto a new gelatin-coated dish to expand the number of cells.

3.3 Retroviral Construct

3.3.1 Insertion of MYOD1 into Retroviral Vector

1. Amplify the coding region of *MYOD1* (NM_002478) by PCR. In detail, 80 ng of *MYOD1* is used as a template for PCR with 5 pmol forward and reverse primers, 12.5 μL of 2× CloneAmp HiFi PCR premix, and sterilized water up to 25 μL. The cycling conditions are 35 cycles of 98 °C for 10 s, 55 °C for 10 s, and 72 °C for 10 s.
2. Detect a single band of 963 bp size by 0.7% agarose gel electrophoresis using 1 μL of the amplified PCR product to confirm that the *MYOD1* sequence is amplified successfully.
3. Clean the PCR product using NucleoSpin Gel and PCR clean-up kit and determine its concentration with NanoDrop.
4. Incubate the mixture containing 2 μL of 10× K buffer, 2 μL of 500 ng/μL pRetroX-TetOne-Puro Vector retroviral vector, 1 μL each of BamH1 and EcoR1, and 14 μL of sterilized water at 37 °C overnight.
5. Detect a single band by 0.7% agarose gel electrophoresis using 1 μL of the digested product to confirm that the retroviral vector was digested successfully.

6. Clean the digested product using NucleoSpin Gel and the PCR clean-up kit and determine its concentration by NanoDrop.
7. To clone the amplified *MYOD1* fragment into the digested retroviral vector, perform an in-fusion cloning reaction. Set up the reaction as 100 ng of purified *MYOD1* fragment, 100 ng of digested retroviral vector, 4 μL of 5× in-fusion HD enzyme premix and sterilized water up to 20 μL, incubate the reaction mixture for 15 min at 50 °C, and then place it on ice.
8. Perform transformation using *E. coli* DH5α competent cells according to the manufacturer's instructions.
9. Pick up the selected colony on LB culture plates and culture those in LB culture medium at 200 rpm at 37 °C overnight.
10. Purify the *MYOD1*-inserted retroviral vector using a HiSpeed plasmid purification kit and quantify its amount using NanoDrop.
11. Confirm that *MYOD1* is correctly inserted into the retroviral vector by direct sequencing of the PCR product amplified by the primer pairs.

3.3.2 Production of Retrovirus

1. For retroviral production, seed GP2–293 packaging cells at 50,000 cells/cm^2 on 100 mm/collagen-coated dish and culture in DMEM with 10% FBS incubated at 37 °C and 5% CO_2 for 24 h (*see* **Note 6**).
2. When GP2–293 packaging cells reach 80% confluency, mix 30 μg of *MYOD1*-inserted retroviral vectors, 30 μg of pVZV-G capsid vectors, and Xfect transfection reagent by vortexing, and incubate those for 10 min at room temperature.
3. Add the mixture to the medium of GP2–293 packaging cells and incubate the culture at 37 °C and 5% CO_2.
4. Collect the viral supernatant and replace with fresh medium at 24 and 48 h after co-transfection.
5. To concentrate the viral supernatant, mix it with Retro-X concentrator, incubate it at 37 °C overnight, and then centrifuge it at 1500× *g* for 45 min at 4 °C.
6. Filter the retrovirus supernatant through a MILTEX HV 0.45 μm filter.
7. Check the titer of the retroviral vector using both One Step TB Green PrimeScript RT-PCR kit and retrovirus titer set with a thermal cycler system according to the manufacturer's instructions.
8. Aliquot the viral supernatant into small aliquots and store them at −80 °C.

3.4 Infection with MYOD1-Retroviral Vector in CD90-Positive UDCs

1. Seed the CD90-positive UDCs at 3000 cells/cm^2 on a gelatin-coated 100 mm dish.
2. After 24 h of seeding, infect the thawed retrovirus at a multiplicity of infection (MOI) of 200 (*see* **Note 7**) with 8 μg/mL hexadimethrine bromide.
3. After 24 h incubation at 37 °C and 5% CO_2, replace the culture medium with fresh GM containing 1 μg/mL puromycin to select the successful *MYOD1*-transduced cells. Change the medium every other day.
4. MYOD1-UDCs should be selected within 7 days after adding puromycin (*see* **Note 8**).

3.5 Myogenic Differentiation of CD90-Positive MYOD1-UDCs

1. Plate CD90-positive MYOD1-UDCs in the collagen-coated wells at a density of 3.5 × 10^4 cells/cm^2. Culture in GM humidified at 37 °C and 5% CO_2.
2. After 24 h, change GM to DM added 1% doxycycline (Dox).
3. After 3 days, replace DM with fresh DM added only 1% Dox. Then, change the medium every 3 days (*see* **Notes 8** and **9**).

3.6 Transfection of ASO into CD90-Positive MYOD1-UDCs

1. Mix ASO with 6 μM Endo-Porter solution and DM to a final concentration of 1–10 μM.
2. Transfect the mixture above into CD90-positive MYOD1-UDCs on day 7 after switching GM to DM. Culture incubated at 37 °C and 5% CO_2.
3. After 72 h of incubation with ASO, change the medium to fresh DM containing 1% doxycycline without ASO.
4. 72 h later, change the medium to fresh DM containing 1% Dox, again.
5. 24 h later, proceed to evaluation (*see* **Note 10**).

3.7 Evaluation of Exon-Skipping Efficiency by RT-PCR

1. Remove DM and wash 1× with PBS. Add cell lysis buffer, lyse the differentiated CD90-positive MYOD1-UDCs, and harvest the total RNA using RNeasy kit.
2. Measure the RNA concentration with NanoDrop.
3. Mix the required reagents for one-step RT-PCR reaction in PCR tubes according to the manufacturer's instructions.
4. Place the PCR tubes with the mixture on a thermocycler. The cycling conditions are 1 cycle 50 °C for 30 min, 1 cycle 95 °C for 15 min, 35 cycles 94 °C for 1 min, 60 °C for 1 min, 72 °C for 1 min, 1 cycle 72 °C for 7 min, and finally hold at 4 °C.
5. Perform microchip electrophoresis and calculate the exon-skipping efficiency using the equation below:
6. Exon-skipping efficiency (%) = [skipped band / (skipped band + non-skipped band)] × 100.

3.8 Detection of Dystrophin by Immunoblotting

1. Extract the total protein of the differentiated CD90-positive MYOD1-UDCs using RIPA buffer containing cOmplete Protease Inhibitor Cocktail.
2. Sonicate the lysates on ice and centrifuge at 14,000× *g* for 15 min at 4 °C.
3. Collect the supernatant and determine protein concentrations using the Pierce BCA protein assay kit.
4. Add 15 μg of total protein in a 0.5 mL tube and dilute by adding RIPA buffer containing cOmplete Protease Inhibitor Cocktail to a total volume of 10 μL.
5. Add 5 μL of NuPAGE LDS sample buffer (4×), 2 μL of NuPAGE sample reducing agent (10x), and 3 μL of deionized water. Denature the cell lysates at 70 °C for 10 min.
6. Prepare NuPAGE Tris–acetate SDS running buffer.
7. Load 20 μL of the sample per lane onto NuPAGE 3–8% Tris–acetate protein gel and perform electrophoresis at 150 V for 75 min with XCell SureLock Mini-Cell.
8. During electrophoresis, soak PVDF for 20 s in methanol and then in EzFastBlot HMW until use (at least 10 min).
9. Cut extra-thick blot filter papers to the same size as that of PVDF and soak those in EzFastBlot HMW until use.
10. After electrophoresis, cut the gel to the same size as that of the PVDF membrane, and soak the gel in distilled water.
11. Place extra thick blot filter paper, PVDF, and gel on the semi-dry transfer apparatus. Transfer at 4 mA/cm^2 for 30 min.
12. Block the transferred PVDF with 5% ECL prime blocking agent.
13. Incubate the blocked PVDF with mouse anti-dystrophin (1: 20) and anti-α-tubulin (1:1200) antibodies diluted by PBS-T respectively, wash with PBS-T repeatedly, and then incubate with anti-mouse horseradish peroxidase-conjugated secondary antibody (1:50,000) diluted by PBS-T.
14. Rinse the reacted PVDF in PBS-T.
15. Detect the proteins using ECL prime western blotting detection reagent and ChemiDoc MP imaging system.

3.9 Detection of Dystrophin by Immunocytochemistry (See Fig. 3)

1. Wash differentiated CD90-positive MYOD1-UDCs with PBS and fix them in 4% formaldehyde for 10 min at 4 °C.
2. Permeabilize those cells in 0.1% nonionic detergent for 10 min at room temperature and block those with 10% GS for 15 min at 37 °C.
3. Incubate those cells with mouse anti-dystrophin (1:20) diluted by 10% GS overnight at 4 °C.

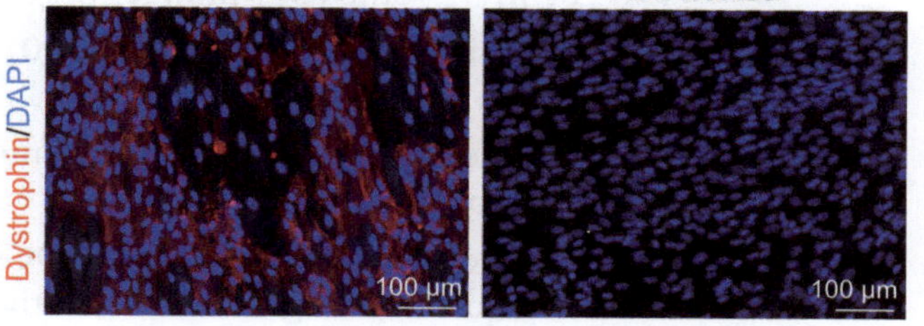

Fig. 3 Representative images of immunocytochemistry in CD90-positive UDC-derived myotubes before/after antisense oligonucleotide (ASO) treatment

4. Wash the cells with PBS and incubate those with Alexa Fluor 546 goat anti-mouse IgG (1:300) diluted by 10% GS for 30 min at room temperature.
5. Wash the cells with PBS and incubate those with DAPI (1: 20,000) diluted by PBS for 10 min at room temperature.
6. Image the cells using BZ-X810 and use BZ-810 Analyzer to semiquantify the fluorescence signal detected at each well of the 96-well plate.

4 Notes

1. Midstream urine is desirable to reduce the risk of bacterial contamination. It is recommended that the collected urine is used for isolation within 4 h and stored at room temperature. Even keeping the cells at 4 °C might reduce the cell viability.
2. The fluid level of 3 mL is as high as the point around the beginning of the curve in the 50 mL tube.
3. The cell pellet of UDCs during isolation is sometimes invisible. Therefore, leaving 3 mL of urine is important to avoid discarding the cell component.
4. Colonies of UDCs appear within 1–2 weeks. To find the colony using microscopy, it is better to check from the edge of the well clockwise. In our experience, the first colony of UDCs tends to appear at the edge of the well.
5. The proportion of CD90-positive UDCs varies depending on the sample. Since the proportion is different at every urine collection from the same donor, it is worth re-isolating UDCs to obtain CD90-positive UDC-dominant urine samples.
6. Collagen coating is necessary for culturing HEK293 packaging cells. Transfection may be better at 24 h or more after seeding because packaging cells easily detach from the culture plate.
7. We usually calculate MOI by the equation below:

8. MOI = the copy number of retrovirus/the number of UDCs.
9. In this protocol, MOI is 200, and the number of UDCs is 1.6×10^5 cells/100 mm dish.
10. Therefore, the required copy number of retrovirus is 3.2×10^7 copies.
11. The *MYOD1*-transduced UDCs could be stored at −80 °C or liquid nitrogen before the induction of myogenic differentiation.
12. We usually add Dox to DM just before each experiment.
13. UDCs fuse and form myotubes within 2–3 weeks after switching GM to DM.

Acknowledgments

This work was supported by the Japan Society for the Promotion of Science Grants-in-Aid for Scientific Research (C) (grant number 18 K075449), Grants-in-Aid for Research on Nervous and Mental Disorders (grant numbers 28-6 and 30-9), and the Japan Agency for Medical Research and Development (grant numbers 18ek0109239h0002, 18lm0203066h0001, and 18lm0203069h0001).

Conflict of Interest The authors have no conflicts of interest to declare that are relevant to the content of this chapter.

References

1. Hoffman EP et al (1987) Dystrophin: the protein product of the Duchenne muscular dystrophy locus. Cell 51:919–928
2. Cirak S et al (2011) Exon skipping and dystrophin restoration in patients with Duchenne muscular dystrophy after systemic phosphorodiamidate morpholino oligomer treatment: an open-label, phase 2, dose-escalation study. Lancet 378:595–605
3. Komaki H et al (2018) Systemic administration of the antisense oligonucleotide NS-065/NCNP-01 for skipping of exon 53 in patients with Duchenne muscular dystrophy. Sci Transl Med 10(437):eaan0713
4. Antoury L et al (2018) Analysis of extracellular mRNA in human urine reveals splice variant biomarkers of muscular dystrophies. Nat Commun 9:3906
5. Takizawa H et al (2019) Modelling Duchenne muscular dystrophy in MYOD1-converted urine-derived cells treated with 3-deazaneplanocin a hydrochloride. Sci Rep 9:3807
6. Kunitake K et al (2024) Characterization of CD90/Thy-1 as a crucial molecular signature for myogenic differentiation in human urine-derived cells through single-cell RNA sequencing. Sci Rep 14(1):2329

Chapter 5

Lipidomic Sample Preparation to Detect Phosphatidylcholine Alterations in Skeletal Muscle

William J. Valentine, Suzumi M. Tokuoka, Yoshihiro Kita, and Hideo Shindou

Abstract

Altered phospholipid compositions in skeletal muscle occur in muscular dystrophy and other pathological conditions; however, origins and relationships to the disease course are unclear. Liquid chromatography–mass spectrometry (LC–MS) technologies allow the determination of phospholipid compositions and alteration patterns in disease states. Here, we describe a basic protocol to prepare methanolic extracts from skeletal muscle, which may be analyzed by lipidomic LC–MS in order to detect compositional alterations in phosphatidylcholine, a major phospholipid species of skeletal muscle, as well as other phospholipid classes such as phosphatidylethanolamine. These analyses are useful to determine relative increases or decreases in individual phospholipid species between samples and capture changes in membrane compositions rather than the absolute quantities of lipids. LC–MS lipidomic analyses often yield large datasets containing hundreds of chromatographic peaks for each sample, and a ratiometric analysis strategy to determine the relative abundances of the major PC species is also briefly introduced.

Key words Phosphatidylcholine, Skeletal muscle, Lipidomic analyses, Phospholipid compositions, Muscular dystrophy

1 Introduction

Cellular membranes are composed mainly of lipids and proteins, and phospholipids, along with cholesterol and sphingolipids, are the primary lipid components. The amphipathic structure of phospholipids supports the formation of lipid bilayers with myriad cellular functions such as separation of the cell from its environment, formation and delineation of intracellular organelles and compartments, as platforms for molecular interactions [1], and as storage of precursors for bioactive lipid mediators [2].

A great diversity of phospholipid species occurs due to the combinatorial possibilities of the choice of polar headgroups combined with the choices of two fatty chains. The fatty chains found in

Yoshitsugu Aoki and Norio Motohashi (eds.), *Splice Switching: Methods and Protocols*, Methods in Molecular Biology, vol. 2963, https://doi.org/10.1007/978-1-0716-4738-7_5,

the various classes of phospholipids are most commonly 12 to 24 carbons in length and contain between 0 and 6 double bonds. The choice of headgroups defines the glycerophospholipid classes phosphatidic acid (PA), phosphatidylcholine (PC), phosphatidylethanolamine (PE), phosphatidylserine (PS), phosphatidylinositol (PI), and phosphatidylglycerol (PG), respectively, as well as the structurally unique phospholipid cardiolipin (CL) with its unique diphosphatidyl, PG-based structure [3]. Among these classes, PC followed by PE is generally most abundant in mammalian cell membranes, together accounting for the majority of the phospholipids in the membranes.

The fatty chain compositions of phospholipids are likely to play roles in various muscle diseases. The source and impacts of altered phospholipid compositions are an area of active research and are expected to lead to new therapeutic strategies to lessen the severity of muscle diseases and improve muscle health. For instance, in Duchenne muscular dystrophy (DMD), the sarcolemma that surrounds muscle fibers is susceptible to contraction-induced damage and allows unregulated passage of ions and other charged molecules, leading to myofiber degeneration and progressive muscle wasting. PC is a main component of the sarcolemma that surrounds muscle fibers, and the fatty chain compositions of PC have been long-noted to be altered in the DMD muscle [4]. PE fatty chain compositions may also impact muscle diseases and are suggested to impact muscle atrophy [5]. In this protocol, we describe a basic protocol to prepare methanolic extracts from skeletal muscle which may be analyzed by lipidomic LC–MS in order to detect compositional alterations in PC and PE. It can be used to determine relative increases or decreases in individual phospholipid species between samples and captures changes in membrane compositions rather than absolute quantities of lipids. LC–MS lipidomic analyses often yield large datasets containing hundreds of chromatographic peaks for each sample, and a ratiometric analysis strategy that determines relative abundances of the major PC and PE peaks is also described.

2 Materials

All solutions should be prepared using ultrapure-grade water and reagents.

Organic solvents should be of HPLC- or LCMS-grade purity.

2.1 Reagents and Supplies

1. Liquid nitrogen.
2. 2 mL crushing tubes with stainless steel inner lids (BMS, catalog number: MT020-01HS).
3. Metal crushers (stainless steel bullets) (Tokken, Inc., catalog number: SK-100-DLC10).

4. Methanol (HPLC-grade) (Fujifilm Wako, catalog number: 132–06471).
5. Ultra-pure water (MilliQ).
6. Sucrose (Fujifilm Wako, catalog number: 195–07925).
7. Metal tongs.
8. Styrofoam coolers.

2.2 Instruments

1. Surgical instruments for mouse muscle dissection.
2. Cryostat (Thermo Scientific CryoStar NX70).
3. Automill cryogenic pulverizer (Tokken, Inc., model: Automill TK-AM5).
4. Sample rack for Automill (24 × 2 mL tubes) (Tokken, Inc., catalog number: TK-AM7–24).
5. Laboratory rotator (Taitec, model: RT-50).

3 Methods

In measurements of phospholipids using our method, where the objective is the relative quantification of PC and PE between control and experimental samples (cell culture or tissues from experimental animals), extraction with methanol as described here is a preferred method. This method is simple and has a low risk of procedural errors. The key points are to achieve complete powderization of samples kept constantly in the frozen state and to use a suitable amount of tissue for this purpose. Suitable amounts of fresh muscle tissue for pulverization (~20–50 mg) may be directly frozen in 2 mL pulverizing tubes and stored frozen and ready for pulverization, as described in Subheading 3.3. Alternatively, suitable amounts of cryosectioned frozen tissue can be obtained from frozen whole muscles (described in Subheading 3.2).

3.1 Tissue Collection

1. Mice are euthanized by isoflurane inhalation, followed by cervical dislocation.
2. Hindlimb muscles such as tibialis anterior, gastrocnemius, extensor digitorum longus, and soleus are obtained rinsed in PBS and stored on ice before freezing.
3. Whole muscles may be snap-frozen in liquid nitrogen-chilled isopentane according to standard protocols for the processing of fresh frozen muscles for histological analyses. In our experience, freezing tissues in isopentane does not interfere with lipidomic analyses.

4. If histological analyses are not required, whole muscles or muscle pieces may be snap-frozen by placing tissues in a microfuge tube and dropping the tissues in liquid nitrogen.
5. For muscle pieces, appropriately sized samples (20–50 mg) from midbelly regions or other regions of interest are placed in 2 mL crusher tubes and dropped in liquid nitrogen for flash freezing.
6. The frozen muscle samples may be stored at -80 °C before cryosectioning and/or crushing to obtain suitable samples for LCMS.

3.2 Collecting Cryosections from Frozen Muscle for LC–MS Analyses

This section describes how to collect suitable tissue samples for crushing and lipid extraction using fresh frozen whole muscles as the starting material. Cryosectioning allows suitable amounts of frozen tissue for pulverization from the region of interest of the muscle and affords the option to collect additional sections/tissue samples from the same muscle which can be used for complementary analyses of muscle histology, gene expression, etc. Generally, we obtain cryosections from the middle third of the muscle near the midbelly to obtain representative samples from the frozen muscles, and other regions may also be obtained depending on the purpose.

1. Prepare a cryostat instrument by cooling down the specimen head, blade holder, and chamber to appropriate temperatures for sectioning muscle.
2. Remove samples from the freezer and place on dry ice. Keep samples frozen during all stages of cryosectioning and crushing prior to lipid extraction.
3. Chill a sample chuck in the cryostat chamber. Adhere the muscles to the chuck by applying ice cold 20% sucrose to the prechilled chuck and rapidly freezing the muscle to the chuck (*see* **Note 1**). The key is to have the sample chuck very cold; apply ~100–200 μL of 20% sucrose solution, and, grasping the frozen muscle near the midbelly region with chilled tweezers, insert the distal portion of the muscle into the sucrose drop just as it freezes. The muscle should be frozen to the chuck in an upright position suitable for cutting transverse sections (*see* **Note 2**).
4. After the sucrose solution freezes, apply ~100–200 μL more to reinforce the frozen connection of the muscle to the chuck. Repeat this procedure several times in order to thoroughly adhere the muscle to the sample chuck. Leave about ~70% of the muscle exposed for sectioning and not covered with sucrose solution (*see* **Note 3**).
5. Collect ~20–50 mg tissue/sample (i.e., approximately thirty 20 μm slices of a TA muscle from an adult mouse) and transfer

to a chilled 2 mL crusher tube placed inside the cryostat staging area using chilled tweezers. Keep samples frozen at all times and store at -80 °C until further processing.

3.3 Tissue Pulverization with Automill

1. Samples are kept frozen before crushing and lipid extraction. If tissue samples are not already in the 2 mL pulverizing tubes, as detailed in Subheadings 3.1 and 3.2, they may be placed in the 2 mL pulverizing tubes just before crushing (step 4 in this list). Both the metal crusher and 2 mL pulverizing tube are chilled to a freezing temperature before they contact the tissue (to prevent the tissue from thawing), as detailed in the following steps.
2. Place a cardboard 81-place microcentrifuge tube storage box, with the cardboard box cover removed, in a styrofoam cooler. Pour enough liquid nitrogen into the tube storage box to fill it about half way, and pour additional liquid nitrogen into the styrofoam cooler so that the liquid nitrogen level is uniform inside and outside the cardboard box (see Fig. 1).
3. Place the 2 mL pulverizing tubes (either with or without samples), with their screw cap lids removed, in the cardboard box so that they are partially submerged in liquid nitrogen. The liquid nitrogen level should be low enough so that as the tubes float slightly in liquid nitrogen they remain in place and upright in the cardboard tube storage box (see Fig. 1).
4. If not already in the tube, place the frozen tissue to be pulverized (~20–50 mg) into the 2 mL pulverizing tubes, using

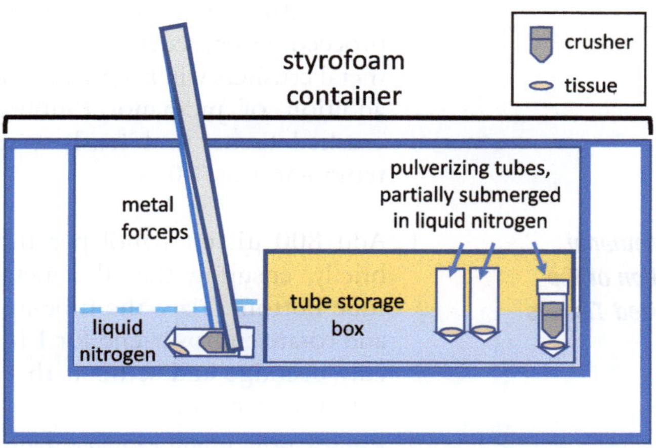

Fig. 1 Putting tissues and metal crushers together in the pulverizing tubes with everything cooled. When adding the metal crusher, keep the tube horizontal and maintain this orientation as the screw cap is tightened and the tube is placed in liquid nitrogen, maintaining the position until the metal crusher has had time to chill

chilled tweezers to transfer the tissues. Make sure the tissues fall to the bottom of the tubes.

5. Grasp a tube with the frozen sample (using forceps if necessary) and partially insert a metal crusher while keeping the tube tilted horizontally so that the crusher does not slide to the bottom. Keeping the tube horizontal, screw on the lid, and using large forceps or tongs, place it into liquid nitrogen and hold it in the horizontal position long enough that the metal crusher has time to cool before it will contact the tissue (see Fig. 1).
6. Put the tube back into the cardboard box that is partially submerged in liquid nitrogen.
7. Repeat and add a metal crusher to all of the samples.
8. Pour liquid nitrogen into another styrofoam container, and, using a pair of large metal forceps or tongs, immerse the metal base of the sample rack for the automill in liquid nitrogen to chill it.
9. Remove the base of the sample rack from liquid nitrogen (using metal tongs), and using forceps insert the samples into the prechilled base, apply the metal cover to the base, and tighten the wingnut screws to secure it.
10. Shake the tubes using the Tokken automill (700–1500 rpm, 20 s). Repeat two to three times until fully pulverized. Be careful at this point because tubes may break if the tissue amount is too small or if the shaking speed is too fast, and the samples will be lost.
11. Remove the sample rack from the automill and collect the tubes with the crushed samples. Put samples on ice and proceed with methanol extraction of the pulverized tissues. If proceeding immediately, ice is not needed because the chilled metal crushers will keep samples frozen temporarily during the addition of methanol (Subheading 3.4). The tubes with crushed tissues and crushers still inside may also be stored temporarily at -80 °C.

3.4 Methanol Extraction of the Pulverized Tissues

1. Add 800 μL methanol per tube. Shake or vortex the tube briefly, ensuring that the metal crusher dislodges from the tube bottom. Place the tube in a rotator in a 4 °C cold room and rotate end-over-end for 1 h. The metal crusher will repeatedly dislodge and settle at the other end of the sample tube with each rotation.
2. Remove the metal crushers from the tubes using tweezers and recap the tubes.
3. Spin the tubes at 14,000 rpm for 10 min at 4 °C.

4. Transfer 500 μL of methanolic extract from each tube to a new 1.6 mL microfuge tube. These samples can be stored at -80 °C indefinitely until analysis.
5. It may be beneficial to dilute all extracts for lipidomic analyses to a similar concentration, such as corresponding to ~5 mg of original tissue per mL methanol, for easier final dilution of batches of samples to the appropriate concentrations for lipidomic analyses. These final concentrations vary for different instruments and analytical purposes and often have to be empirically determined. In our experiences, these final concentrations can range from 10–300 μg of original tissue per mL methanol.
6. To further dilute the samples before storage, spin the tubes again at 14,000 rpm for 10 min at 4 °C. Transfer the appropriate amount of supernatant to a new tube and dilute with methanol to the desired concentration (i.e., 5 mg of original tissue/mL) and store at −30 °C (for short term, <2 weeks) or at −80 °C (longer term) before LC–MS analyses.

3.5 Sample Dilution for LC–MS Analyses

1. Remove the samples from the freezer and allow to warm to 4 ° C.
2. Centrifuge at 14,000 rpm for 10 min at 4 °C.
3. Transfer an aliquot of the supernatant to an LC–MS sample tube and dilute with methanol to an appropriate concentration for LC–MS analyses, based on the original tissue weights. In our experience, the appropriate concentration of extracts applied to 8050 LCMS for PC analyses may correspond to 10–300 μg original tissue / mL methanol, for a 5 μL injection volume to the LC system. Ask advice from the LC–MS operators for the appropriate concentrations, and test samples (starting with lower concentrations) may be run to determine the suitable concentrations. Signals for several PC peaks should be detected well above background, while signals of the largest peaks should not saturate the MS detector.
4. The diluted extracts in LC–MS sample tubes may be stored at 4 °C for several days before analysis. The diluted extracts stored for longer or being reanalyzed should be transferred to microcentrifuge tubes, centrifuged (14,000 rpm for 10 min at 4 °C) to pellet any insoluble precipitate, and supernatants transferred to fresh tubes before analysis.

3.6 LC–MS Analysis

Phospholipid analyses of the muscle tissues are carried out in a laboratory facility equipped with specialized equipment and skilled personnel for LC–MS-based lipidomic analyses.

We employ LC–MS methods that contain over 1000 multiple-reaction monitoring (MRM) events which target headgrouprelated phospholipid fragmentation. MRM transitions targeting the headgroups do not distinguish the isomers of the fatty chains, and each chromatographic peak may contain multiple isomers with differing chain sets. This strategy is sufficient for preliminary analyses and some studies, and additional LC–MS–MS strategies can be employed to identify precise isomers when required. For background and more detailed description of lipidomic LCMS theory and methodologies, see refs [6–9]. In the following is briefly described a methodology we employed in some studies [10, 11].

3.6.1 Phospholipid Analysis by LC–MS

For lipidomic analyses, methanolic extracts of muscle tissue samples were further diluted with methanol to a final concentration corresponding to ~150 μg of original wet tissue weight/mL. LC-selected reaction monitoring (SRM)–MS analyses were performed with a Nexera UHPLC system coupled to a LCMS-8050 triple quadrupole mass spectrometer (Shimadzu). An Acquity UPLC BEH C8 column (1.7 μm, 2.1 mm x 100 mm, Waters) was used with the following ternary mobile phase compositions: 5 mM NH_4HCO_3/water (mobile phase A), acetonitrile (mobile phase B), and isopropanol (mobile phase C). The pump gradient [time (%A/%B/%C)] was programmed as follows: 0 min (50/45/5)–10 min (20/75/5)–20 min (20/50/30)–27.5 min (5/5/90)–28.5 min (5/5/90)–28.6 min (50/45/5). The flow rate was 0.35 mL/min, and the column temperature was 47 °C. Sample injection volumes were 5 μL. Comprehensive LC–SRM–MS analysis was performed by positive ion mode electrospray ionization with the transitions $[M + H]^+ \rightarrow 184$ for PC and $[M + H]^+ \rightarrow [M + H\text{-}141]^+$ for PE to detect all diradyl PC and PE species possessing two even number carbon chains, each 14–24 carbons in length.

3.7 Data Analyses

LC–MS data analysis for phospholipids from biological samples requires careful peak picking from each chromatogram. The automatic peak picking functions of analytical software are often not completely accurate with complex phospholipid chromatograms, and manual confirmation of each specimen's chromatogram one by one often becomes necessary. Moreover, ^{13}C isotopes of phospholipids with a lower precursor m/z often appear in the MRM chromatograms of phospholipids with a higher precursor m/z, which potentially leads to errors in the analysis. For example, PC 38:6 isotopes containing two ^{13}C appear in the same MRM transitions that detect monoisotopic PC 38:5. For LC–MS data analysis, we use TRACES, a lightweight chromatogram browser for MRM-based targeted LC–MS analysis [10]. This software enables calculation of any target peak area value by defining the MRM transition (m/z for Q1 and Q3) and the area under the curve integration range (start and end times). The isotope removal function of this software prevents inappropriate peak analysis for biological samples containing complex mixtures of molecules with varying levels of fatty acyl unsaturation.

In the next section, we briefly list the steps for the calculation and graphing of PC major peak area values. The peak areas were obtained and deisotoped using TRACES software [12], and values were inserted into a Microsoft Excel spreadsheet for further calculations. The calculation steps are described as they are performed for each sample. Performing these calculations in batch procedures in Excel spreadsheets is an efficient way to process data for the larger numbers of samples typically analyzed in lipidomic analyses, and these steps are equally suited to analyze any class of phospholipid (PC, PE, PS, etc.).

3.7.1 Steps for the Calculation of PC Major Peaks

For each sample:

1. The peak areas of all PC peaks are summed to obtain the total peak areas for PC signals.
2. The peak areas of each individual PC peak are normalized to the sum, and its value is expressed as percentage of the total PC. The calculation for each peak is: % of total PC = (individual peak area / total peak area) × 100.
3. Peaks with low signal similar to background in all samples are considered as noise and omitted from the analysis. We often employ a 1% threshold, and PC peaks with values that are less than 1% of the total PC in every sample are assumed to represent background. In this protocol, we use a 1% threshold as an example; however, in deciding the threshold, one should consider the peak appearance and overall noise levels, as well as the purpose of the analysis.
4. Peaks which are less than 1% of the total in all samples are assumed to represent background and are deleted from the analyses in all samples. However, if a peak is at least 1% in ANY sample, it is retained for further calculations in ALL samples. It is critical to retain the same set of peaks for analysis in all samples being compared in an analysis.
5. Make a copy of the original spreadsheet, and delete all the background peaks from the original.
6. Recalculate the summed total PC signal.
7. Normalize the individual PC species to this new sum, and express the peak values as the percentage of the total.

3.7.2 Example Strategy to Plot the Most Abundant Peaks

For each analysis, the most abundant peaks may be displayed in a graph by specifying that all peaks that are of at least a specified percentage of the total PC or PE signals, in at least one sample, are plotted. Figure 2 shows an example of graphed data of PC compositions in mouse skeletal muscles, in which all PC peaks with area

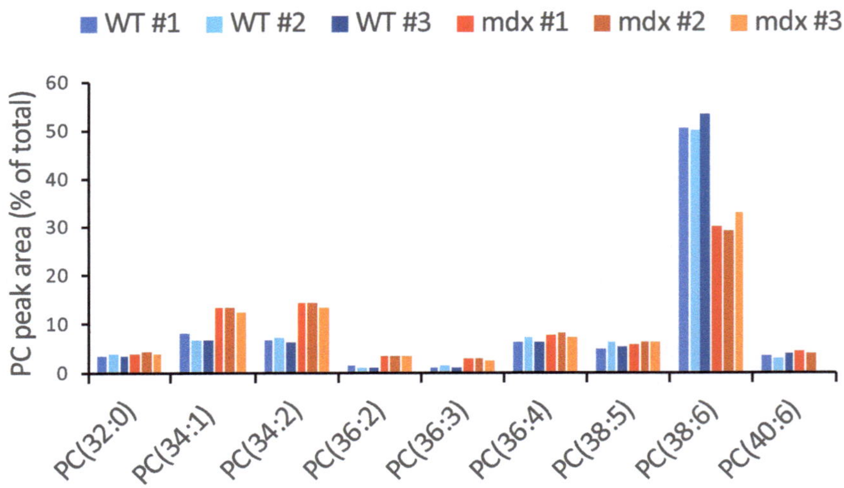

Fig. 2 PC profiles of gastrocnemius of three wild-type (WT) and three mdx mice (12-week-old male mice). The most abundant PC peaks, comprising at least 3% of the total PC in any sample, were plotted

values greater than or equal to 3% of the total PC peak area values, in at least one sample, are plotted. By using a lower (i.e., 2%)–higher cutoff (i.e., 5%), the number of peaks to be displayed on a graph may be increased or decreased, respectively. Since less abundant peaks may also be of great interest, the data presentation should be adapted to best suit the information to be conveyed.

4 Notes

1. In our experience, 20% sucrose freezes the sample to the chuck more stably and cracks less often than frozen water. We avoid the use of OCT or tragacanth gum on lipidomic samples because of the extreme sensitivity of the assays.
2. A small amount of thawing of the muscle when it initially contacts sucrose solution is unavoidable. This is minimized by keeping the sucrose solution chilled on ice and precooling the sample chuck as much as possible.
3. We avoid contacting the frozen sucrose with the cryostat blade as much as possible, but small amounts of sucrose collected with the tissues do not interfere with the most downstream analyses.

Acknowledgments

This work was supported by the Mishima Kaiun Memorial Foundation to WJV, the National Center for Global Health and Medicine, Intramural Research Fund (22 T001 to HS), the Japan Health

Research Promotion Bureau Research Fund (JH2022-B-03 to HS), and the TERUMO LIFE SCIENCE FOUNDATION (22-III4054 to HS).

Department of Lipid Life Science, Japan Institute for Health Security collaborates with ONO PHARMACEUTICAL Co., Ltd. (Osaka, Japan) and Shimadzu Corp. (Kyoto, Japan), although both companies did not financially support this manuscript.

Conflict of Interest The authors have no conflicts of interest to declare that are relevant to the content of this chapter.

References

1. van Meer G, Voelker DR, Feigenson GW (2008) Membrane lipids: where they are and how they behave. Nat Rev Mol Cell Biol 9(2): 112–124
2. Shimizu T (2009) Lipid mediators in health and disease: enzymes and receptors as therapeutic targets for the regulation of immunity and inflammation. Annu Rev Pharmacol Toxicol 49:123–150
3. Valentine WJ et al (2022) Update and nomenclature proposal for mammalian lysophospholipid acyltransferases, which create membrane phospholipid diversity. J Biol Chem 298(1): 101470
4. Takagi A et al (1968) Fatty acid composition of lecithin from muscles in human progressive muscular dystrophy. Clin Chim Acta 20(1): 41–42
5. Eshima H et al (2023) Lipid hydroperoxides promote sarcopenia through carbonyl stress. elife 12:12
6. Taguchi R et al (2005) Focused lipidomics by tandem mass spectrometry. J Chromatogr B Analyt Technol Biomed Life Sci 823(1):26–36
7. Taguchi R, Nishijima M, Shimizu T (2007) Basic analytical systems for lipidomics by mass spectrometry in Japan. Methods Enzymol 432: 185–211
8. Fahy E et al (2007) Bioinformatics for lipidomics. Methods Enzymol 432:247–273
9. Ivanova PT et al (2007) Glycerophospholipid identification and quantitation by electrospray ionization mass spectrometry. Methods Enzymol 432:21–57
10. Hashidate-Yoshida T et al (2015) Fatty acid remodeling by LPCAT3 enriches arachidonate in phospholipid membranes and regulates triglyceride transport. elife 4:4
11. Valentine WJ et al (2021) Lipidomic analyses reveal specific alterations of phosphatidylcholine in dystrophic mdx muscle. Front Physiol 12:698166
12. Kita Y et al (2022) TRACES: a lightweight browser for liquid chromatography-multiple reaction monitoring-mass spectrometry chromatograms. Meta 12(4)

Chapter 6

Evaluation for the Efficacy of Exon-Skipping Therapy Using In Vivo Assessment of Skeletal Muscle Function and Fragility

Yuki Ashida, Katsura Minegishi, Norio Motohashi, Yoshitsugu Aoki, and Takashi Yamada

Abstract

Splice-switching oligonucleotide (SSO) therapy is expected to be a promising treatment for muscular diseases such as Duchenne muscular dystrophy (DMD). Preclinical studies using animal models of muscle diseases are essential for the development of SSO therapy. In addition to the histological and biochemical analyses, an in vivo muscle torque measurement using supramaximal electrical stimulation has become a standard protocol for the evaluation of drug efficacy. Nevertheless, in order to accurately measure the torque production in animal models, there are some critical points that should be brought to attention. In this chapter, we describe detailed methods for in vivo muscle torque measurement in animal models for DMD, i.e., mdx52 mouse and canine X-linked muscular dystrophy in Japan ($CXMD_J$), that set the standard for reproducibility in this field.

Key words Muscle function, Isometric contraction, Eccentric contraction, Prolonged force depression

1 Introduction

The patients with Duchenne muscular dystrophy (DMD) exhibit progressive muscle weakness that eventually leads to loss of ambulation [1]. Various promising therapeutic approaches have been proposed for the treatment of DMD, including genetic and exon-skipping therapy, that aim to restore the expression levels of dystrophin protein [2]. To develop potential therapies, preclinical studies using animal models for muscle diseases are essential. In addition to the histological and biological analyses, measurement of in vivo muscle torque, which enables the assessment of muscle function in living subjects over time, is crucial for the evaluation of drug efficacy. However, there are some critical points that should be brought to attention for measuring the muscle torque precisely.

Yoshitsugu Aoki and Norio Motohashi (eds.), *Splice Switching: Methods and Protocols*, Methods in Molecular Biology, vol. 2963, https://doi.org/10.1007/978-1-0716-4738-7_6,

In this chapter, we will describe a detailed method of in vivo procedures for measuring muscle contractility in wild-type and dystrophin-deficient animals, such as mdx52 mouse and canine X-linked muscular dystrophy in Japan ($CXMD_J$).

1.1 In Vivo Measurement of Isometric Torque

Isometric contraction (ISO), in which the muscle contracts while its length remains unchanged, represents the most popular measure of skeletal muscle function [3]. However, when muscles are electrically stimulated, the response is graded; muscle torque increases with the increased intensity of the stimulus due to the increased number of activated fibers. Therefore, in order to obtain reproducible results, it is important to use supramaximal stimuli, which induce contraction of all muscle fibers in the stimulated muscles. Moreover, as the relationship between torque and frequency is sigmoidal, the maximum ISO torque can normally be obtained at a high stimulation frequency [4]. Previous studies have shown that the in vivo maximum ISO torque normalized by muscle weight (i.e., specific torque) is reduced in rodent and canine models for DMD [5, 6], indicating intrinsic contractile dysfunction in dystrophin-deficient muscles. Theoretically, decreased specific torque can be caused by (i) reduced Ca^{2+} release from sarcoplasmic reticulum (SR), (ii) decreased myofibrillar Ca^{2+} sensitivity, and/or (iii) impaired ability of cross-bridges to generate force in skeletal muscle fibers [7]. Among them, reduced SR Ca^{2+} release [8] and impaired force-generating capacity [9] have been reported in the resting state of dystrophin-deficient muscle fibers.

1.2 In Vivo Measurement of Eccentric Contraction-Induced Torque Drop and Prolonged Torque Depression

It is widely known that dystrophin-deficient muscle is highly susceptible to damage induced by eccentric contractions (ECCs), in which the muscle contracts while being stretched [10]. Indeed, the magnitude of the reduction in torque production during repeated maximal ECCs is greater in dystrophic muscle than in normal muscle [5]. Thus, susceptibility to ECC-induced torque drop has become a standard outcome measure in preclinical studies to assess the efficacy of potential therapies for DMD [11]. Importantly, a growing body of evidence has demonstrated that plasmalemmal excitability measured by M-wave amplitude is reduced in mdx mice, but not wild-type mice, during and immediately after maximal ECCs [12]. Accordingly, maximal ECC-induced torque drop in mdx mice has been associated with both neuromuscular transmission failure [13] and an inability to generate/conduct an action potential across the plasmalemma [14].

Additionally, mechanically demanding ECCs induce a prolonged maximum ISO torque depression in skeletal muscle, which lasts for days or even weeks [15, 16]. Notably, compared to the normal muscle, dystrophic muscle displays severe prolonged maximum ISO torque depression after maximal ECCs [5]. The mechanisms underlying prolonged depression in maximum ISO torque after maximal ECCs in dystrophic muscle are not well understood but can be due to both an impaired SR Ca^{2+} release [5] and a loss of myofibrillar force-generating capacity [17]. From a clinical point of view, prolonged torque depression after ECCs can have a large negative impact on the activity of daily living in patients with DMD.

2 Materials

2.1 Equipment for In Vivo Torque Measurement in a Mouse

1. Torque measuring system with a servomotor (S-14154, Takei Scientific Instruments, Tokyo, Japan) (Fig. 1a).
2. Software to control the servomotor (Takei Scientific Instruments, Tokyo, Japan).
3. Power Lab 2/26 (PL2602, ADInstruments, Sydney, Australia).
4. Software for measuring the torque (LabChart 8; ADInsturments, Sydney, Australia).
5. Electrical stimulator (SEN-3301, NIHON KOHDEN, Tokyo, Japan).
6. Vitrode V disposable surface electrode (V-120SK, NIHON KOHDEN, Tokyo, Japan) (Fig. 1b, c).
7. A custom-made pen-shaped electrode (Fig. 3a, b) (*see* **Note 1**).

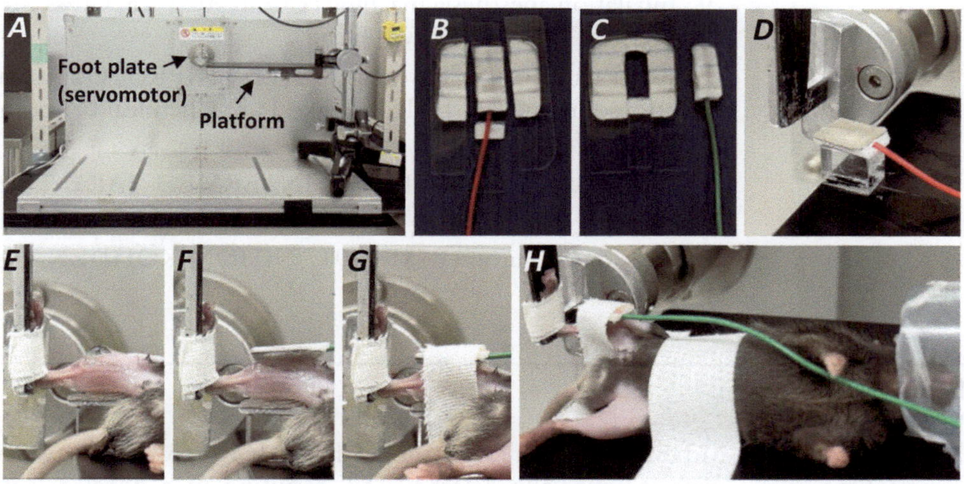

Fig. 1 In vivo torque measurement for plantar flexor muscles in mouse. An overview of the torque measuring system for rodents (**a**). Surface electrodes are trimmed for the posterior (**b**) and anterior (**c**) sides of the lower leg of a mouse. An electrode is pasted to the platform (**d**). The lower leg is put on the electrodes after applying a gel conductor layer to the posterior skin (**e**). The foot is then secured to the footplate of the torque measurement device using an adhesive tape. After applying the gel conductor layer to the anterior skin of the lower leg, another electrode is pasted on the anterior side of the lower leg (**f**) and fixed with an adhesive tape (**g**). The mouse body is secured to the platform using an adhesive tape, taking care not to impede breathing (**h**)

8. Adhesive tape (CH19, NICHIBAN, Tokyo, Japan).
9. Aquasonic 100 ultrasound transmission gel (Parker Laboratories, Inc., NJ, USA).
10. Inhalation anesthesia machine (SN-487-0 T, WAKENYAKU, Kyoto, Japan).
11. Anesthetic agent: isoflurane (FUJIFILM Wako Pure Chemical Corporation, Tokyo, Japan).

2.2 Equipment for In Vivo Torque Measurement in Dog

1. Custom-made torque measuring system with a servomotor (Takei Scientific Instruments, Tokyo, Japan) (Fig. 5a).
2. Software to control the servomotor (Takei Scientific Instruments, Tokyo, Japan).
3. Power Lab 2/26 (PL2602, ADInsturments).
4. Software for measuring the torque (LabChart 8; ADInsturments, Sydney, Australia).
5. Electrical stimulator (SEN-3301, NIHON KOHDEN, Tokyo, Japan).
6. Adjustable operating table for animals with adjustable height (Menix, Saitama, Japan).
7. Adhesive tape (1404BRT, 3 M, Minnesota, USA).
8. Needle electrode (27G, 25 mm, TERUMO, Tokyo, Japan) (Fig. 5b).
9. Inhalation anesthesia machine (COMPACT-15, Cross Medical Service Co., Ltd., Tokyo, Japan).
10. Anesthetic agents: Ravonal (871,115, NIPRO, Osaka, Japan) and isoflurane (FUJIFILM Wako Pure Chemical Corporation, Tokyo, Japan).
11. Intubation tube (971–075, INTERMED JAPAN, Osaka, Japan).
12. Bedside monitor (BSM-2391, NIHON KOHDEN, Tokyo, Japan).

3 Methods

3.1 In Vivo Torque Measurement in a Mouse

Mdx52 mouse is a DMD model in which exon 52 of the dystrophin gene is artificially deleted [18], resulting in a deficiency of the dystrophin protein similar to that seen in DMD patients. These mice exhibit mild motor and myocardial dysfunction, which is characteristic of DMD patients. In addition, mdx52 mice display muscle weakness and hypersusceptibility to ECC-induced muscle damage [5].

3.1.1 Measurement of Isometric Plantar Flexor Torque in a Mouse

1. Anesthetize a mouse with isoflurane inhalation (1.5%, 1.5 L/min) and position it face up under a heating lamp.
2. The skin of the lower leg is shaved, and a gel conductor layer is applied to the posterior skin of the lower leg. The lower leg of the mouse is then put on a surface electrode, which is pasted on the platform (Fig. 1d, e). The foot of the mouse is secured to the footplate of the torque measuring system using an adhesive tape. The height of the platform is adjusted carefully so that the ankle joint of the mouse is set at 0° (i.e., 90° relative to the fibula).
3. After applying the gel conductor layer to the anterior skin of the lower leg, a surface electrode is put on the anterior side of the lower leg and fixed with an adhesive tape (Fig. 1f, g).
4. Fix the body of the mouse to the platform using an adhesive tape, taking care not to impede breathing (Fig. 1h).
5. Perform a 600 ms ISO at 100 Hz stimulation frequency every 1 min adjusting the simulation until a maximal isometric torque is achieved (Fig. 2a) (*see* **Note 2**).
6. Measure the supramaximal electrical stimulation (voltage: ~45 V, pulse duration: 0.5 ms, stimulation duration: 600 ms)-evoked ISO plantar flexor (PF) torque at stimulation frequencies from 1 to 120 Hz. A 1-min rest should be provided between each stimulation to avoid the effects of muscle fatigue (Fig. 2b).

3.1.2 Measurement of Eccentric Contraction-Induced Torque Drop and Prolonged Torque Depression in Mouse Plantar Flexor Muscles

1. Secure a mouse foot to the torque measuring system and measure the ISO PF torque before exposure to ECCs (see Subheading 3.1.1) (*see* **Note 3**).
2. Passive torque is measured during the passive dorsiflexion (without stimulation) of ankle joint from 0° (i.e., 90° relative to the fibula) to 40° at 150°/s.
3. Apply ECC, which comprises forced dorsiflexion from 0° to 40° at 150°/s combined with supramaximal electrical stimulation (voltage: 45 V, frequency: 50–100 Hz, pulse duration: 0.5 ms, stimulation duration: 267 ms) of PF muscles (Fig. 2c) (*see* **Note 4**). Muscles are exposed to 10 repeated ECCs every 4 s (Fig. 2d) (*see* **Note 5**).
4. Measure the ISO PF torque immediately (data not shown) and 48 h after ECCs (Fig. 2e, f).

3.1.3 Measurement of Isometric Dorsiflexor Torque in a Mouse

1. Anesthetize a mouse with isoflurane inhalation (1.5%, 1.5 L/min) and position it face up under a heating lamp.
2. Shave and clean the lateral skin of the mouse lower leg.
3. Secure the foot onto the footplate of the torque measuring system using an adhesive tape. Adjust the height of the

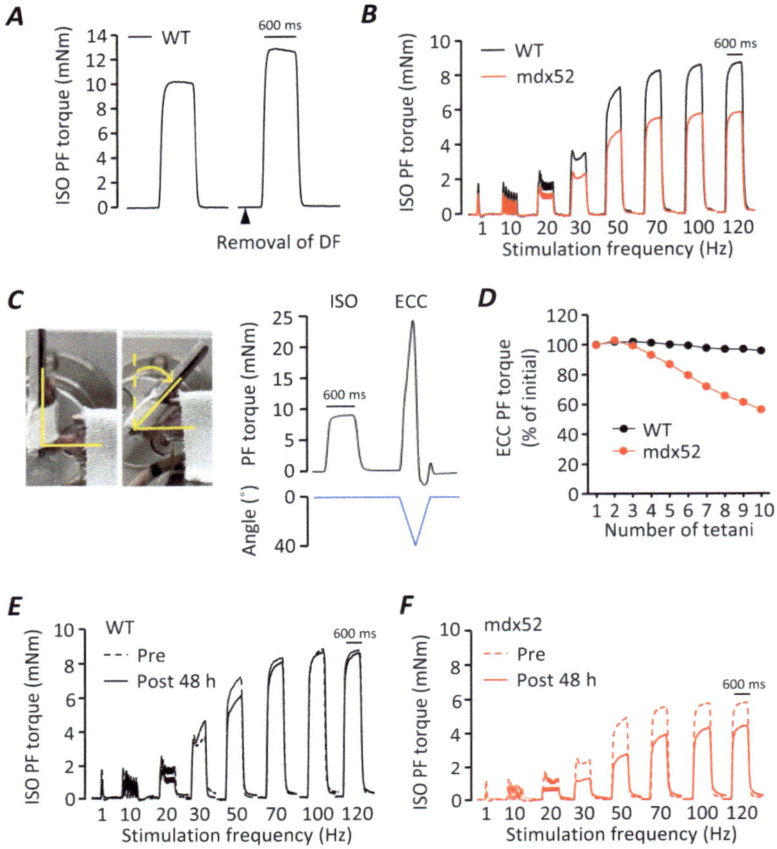

Fig. 2 Typical traces of plantar flexor torque in a mouse. Torque production in isometric contraction (ISO) of plantar flexor (PF) muscles in a wild-type C57BL/6 J mouse (WT, 10 weeks old) (**a**). Note that ISO PF torque is increased after the surgical removal of dorsiflexor (DF) muscles. ISO torque–frequency relationship of PF muscles, with DF muscles intact, in WT (10 weeks old, black line) and mdx52 (9 weeks old, red line) mice (**b**). Comparison between ISO and eccentric contraction (ECC) (0–40° dorsiflexion, 150°/s, 50 Hz, 267 ms) in WT mouse (10 weeks old, black line) (**c**). Active torque during 10 repeated ECCs, which were applied every 4 s (**d**). Data are expressed as a percentage of the initial value. ISO PF torque before (dotted line) and 48 h (solid line) after 10 ECCs in WT (**e**) and mdx52 mice (**f**)

platform so that the ankle joint of the mouse is set at 0° (i.e., 90° relative to the fibula) (Fig. 3c).

4. Apply a gel conductor layer near the fibular head.
5. Place the pen-shaped electrode slightly distal to the fibular head to stimulate the peroneal nerve (Fig. 3d, e). Perform a 600 ms ISO at 100 Hz stimulation frequency every 1 min adjusting the simulation until a maximal isometric torque is achieved (Fig. 4a). The supramaximal stimulation is defined as a

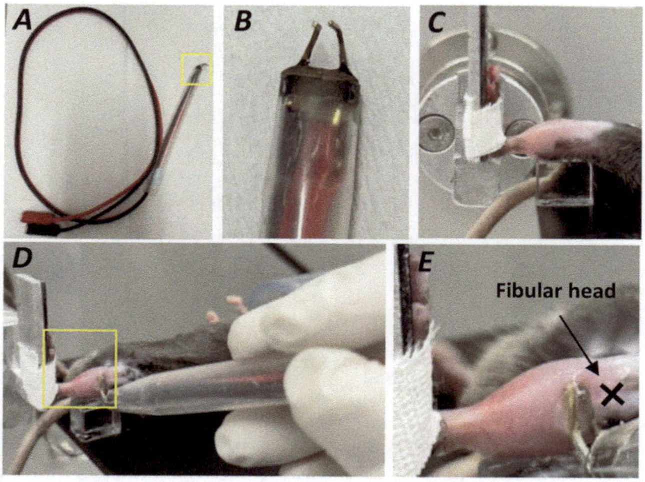

Fig. 3 In vivo torque measurement for dorsiflexor muscles in mouse. Pen-shaped electrode (**a**). An enlarged view of the boxed area in panel A (**b**). The foot of a mouse is secured to the footplate of the torque measuring system using an adhesive tape (**c**). The surface electrode is placed slightly distal to the fibular head to selectively stimulate the peroneal nerve (**d**). An enlarged view of the boxed area in panel D (**e**)

stimulation that has strength significantly higher than the required level to activate all the nerve fibers (*see* **Note 6**).

6. Measure the supramaximal electrical stimulation (voltage: ~25 V, pulse duration: 0.5 ms, stimulation duration: 600 ms)-evoked ISO DF torque at stimulation frequencies from 1 to 120 Hz every 1 min (Fig. 4b).

3.1.4 Measurement of Eccentric Contraction-Induced Prolonged Torque Depression in Mouse Dorsiflexor Muscles

1. Secure a mouse foot to the torque measuring system and measure the ISO DF torque before exposure to ECCs (see Subheading 3.1.3).
2. After applying a gel conductor layer to the posterior and anterior skin of the lower leg, the surface electrodes are fixed with an adhesive tape (see Subheading 3.1.1).
3. Apply ECC, which comprises forced plantar flexion from 0° (i.e., 90° relative to the fibula) to 90° at 270°/s combined with supramaximal electrical stimulation (voltage: ~45 V, frequency: 50–100 Hz, pulse duration: 0.5 ms, stimulation duration: 333 ms) of dorsiflexor (DF) muscles (Fig. 4c) (*see* **Note 7**). Muscles are exposed to ten repeated ECCs every 4 s.
4. Remove the surface electrodes from the lower leg of the mouse and measure the ISO DF torque using pen-shaped electrodes immediately (data not shown) and 48 h after ECCs (Fig. 4d, e).

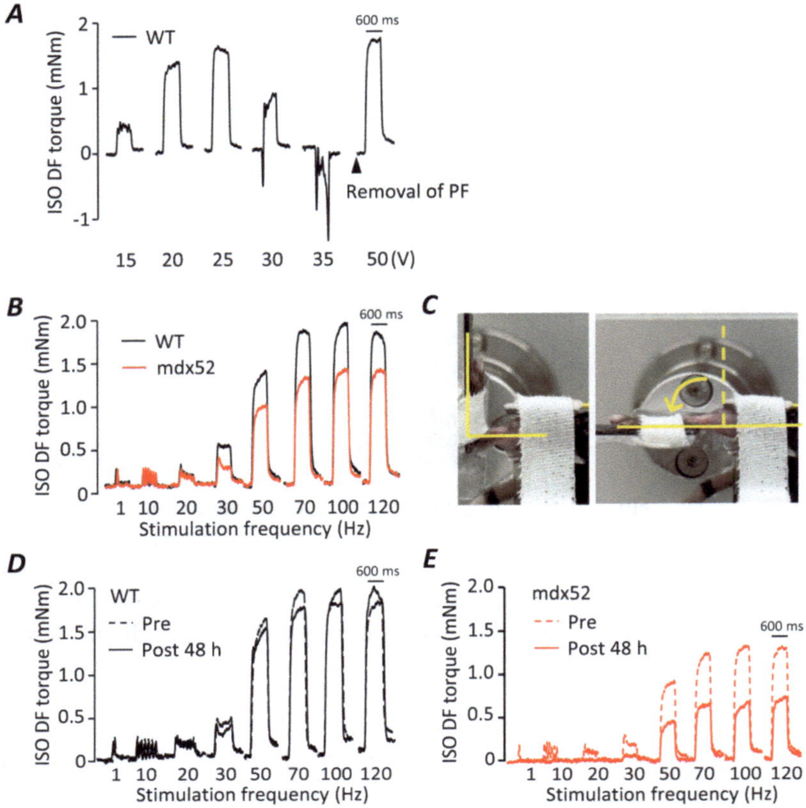

Fig. 4 Typical traces of dorsiflexor torque in a mouse. Torque production in isometric contraction (ISO) of dorsiflexor (DF) muscles during increasing stimulation intensities in a wild-type C57BL/6 J mouse (WT, 13 weeks old) (**a**). Note that plantar flexor (PF) muscles are recruited above 30 V, which masked the DF torque production. On the other hand, the stimulation intensity just below the recruitment of the PF muscles (i.e., 25 V) evokes the maximum DF torque, which is comparable to the maximum DF torque (i.e., 50 V) after surgical removal of the PF muscles. ISO torque–frequency relationship of DF muscles, with PF muscles intact, in WT (13 weeks old, black line) and mdx52 mice (9 weeks old, red line) (**b**). Eccentric contraction (ECC) of DF muscles (0–90° plantar flexion, 270°/s, 50 Hz, 333 ms) (**c**). ISO DF torque before (dotted line) and 48 h (solid line) after 10 ECCs in WT (**d**) and mdx52 mice (**e**)

3.2 In Vivo Torque Measurement in Dog

There is an increasing need for large animal models that present disease progression more similar to that of DMD patients in order to elucidate the pathophysiology and development of drugs. In this context, $CXMD_J$ was produced by artificially inseminating a beagle dog with frozen sperm from a golden retriever with muscular dystrophy (GRMD), which is a dog model of DMD. $CXMD_J$ is characterized by a point mutation in the consensus splicing site of intron 6 of the canine dystrophin gene, resulting in a deficiency of dystrophin [19]. $CXMD_J$ shares common clinical symptoms with DMD patients, including skeletal muscle atrophy and joint contracture. In addition, there is a difference in the fiber types between large mammals (i.e., dogs and humans) and rodents. The former do not have the fastest type IIb fiber [20, 21], which is known to be particularly sensitive to ECCs [22], while it is abundantly observed in the latter [23].

3.2.1 Measurement of Isometric Dorsiflexor Torque in Dog

As has been reported in GRMD [24], dorsiflexion angle is severely limited in $CXMD_J$ because of the tarsal joint contracture. For this reason, it is not feasible to apply ECCs on PF muscles in $CXMD_J$ dogs. Hence, we have developed a method for measuring DF torque in dogs. This system is a modification of the device established by Kornegay et al. [25] for measuring ankle torque in GRMD.

1. After an intravenous administration of thiopental sodium (Ravonal; 20 mg/kg) to a dog, tracheal intubation is performed, and anesthesia is maintained with inhaled isoflurane (2.0–2.5%). During anesthesia, heart and respiratory rates and saturation of hemoglobin by peripheral oxygen (SpO_2) are monitored continuously with a pulse oximeter. The isoflurane dose is adjusted to maintain SpO_2 at 97–100% during sedation.
2. Shave and clean the skin near the fibular head of the dog.
3. The dog is placed in a semi-supine position, and the foot is secured to the footplate of the torque measuring system using an adhesive tape. Adjust the height of the operating table so that the dog ankle joint is set at 0° (i.e., 90° relative to the fibula) and the knee joint is positioned at 90° (Fig. 5c). At this time, care should be taken to avoid inversion and eversion of the ankle joint (Fig. 5d).
4. After disinfecting the skin with alcohol, a pair of needle electrodes are inserted slightly distal (about 10–20 mm) from the fibular head, where the peroneal nerve is located (Fig. 5e, f).
5. Perform a 600 ms ISO at 100 Hz stimulation frequency every 1 min adjusting the simulation until a maximal isometric torque is achieved (Fig. 6a). The supramaximal stimulation is defined as a stimulation that has strength significantly higher than the required level to activate all the nerve fibers (*see* **Note 8**).
6. Measure the supramaximal electrical stimulation (voltage: ~30 V, pulse duration: 0.5 ms, stimulation duration: 600 ms)-evoked ISO DF torque at stimulation frequencies from 1 to 120 Hz every 1 min (Fig. 6b).

3.2.2 Measurement of Eccentric Contraction-Induced Torque Drop and Prolonged Torque Depression in Dog Plantar Flexor Muscles

1. Secure the foot of a dog to the torque measuring system, as described in Subheading 3.2.1, and measure the ISO DF torque before exposure to ECCs.
2. Passive torque is measured during passive plantar flexion of the ankle joint from 0° (i.e., 90° relative to the fibula) to 60° at 290°/s.

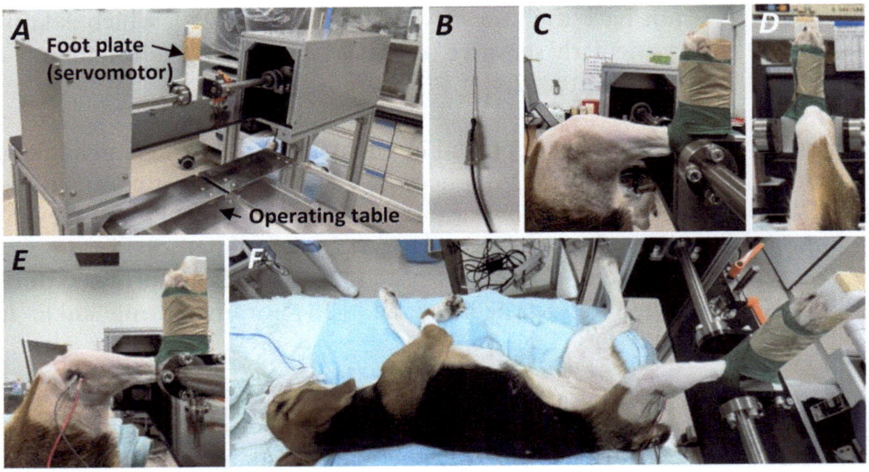

Fig. 5 In vivo torque measurement for dorsiflexor muscles in a dog. An overview of the torque measuring system for dogs (**a**). Needle electrode (27G, 25 mm) (**b**). The foot of a dog is secured to the footplate of torque measurement device using an adhesive tape (**c**). Adjust the height of the operating table so that the dog ankle joint is fixed at 0° (i.e., 90° relative to the fibula) and the knee joint is positioned at 90°. At this time, care should be taken to avoid inversion and eversion of the ankle joint (**d**). A pair of needle electrodes are inserted slightly distal (about 10–20 mm) from the fibular head, where the peroneal nerve is located. The needle electrodes are then fixed with a tape (**e**). Overall view of the dog during torque measurement (**f**)

3. Apply ECC, which comprises forced plantar flexion from 0° to 60° at 290°/s combined with supramaximal electrical stimulation (voltage: 45 V, frequency: 50–100 Hz, pulse duration: 0.5 ms, stimulation duration: 207 ms) of DF muscles (Fig. 6c). Muscles are exposed to 30 repeated ECCs every 4 s (Fig. 6d) (*see* **Note 9**).
4. Measure the ISO DF torque immediately (data not shown) and 48 h after ECCs (Fig. 6e, f) (*see* **Note 10**).

4 Notes

1. The pen-shaped electrode, platform, and mouse footplate were all custom-manufactured by the university machine shop.
2. This way of electrical stimulation activates both the PF muscles and DF muscles. The antagonistic actions of the DFs can counter some of the force produced by the PFs. However, as has been shown in the previous study [26], PF torque was increased only by 10–20% after the surgical removal of the DFs (Fig. 2a). Note that a direct electrical stimulation of tibial nerve also activates DF muscles when using supramaximal stimulation, likely due to the electrical current spread to antagonist muscles [27].

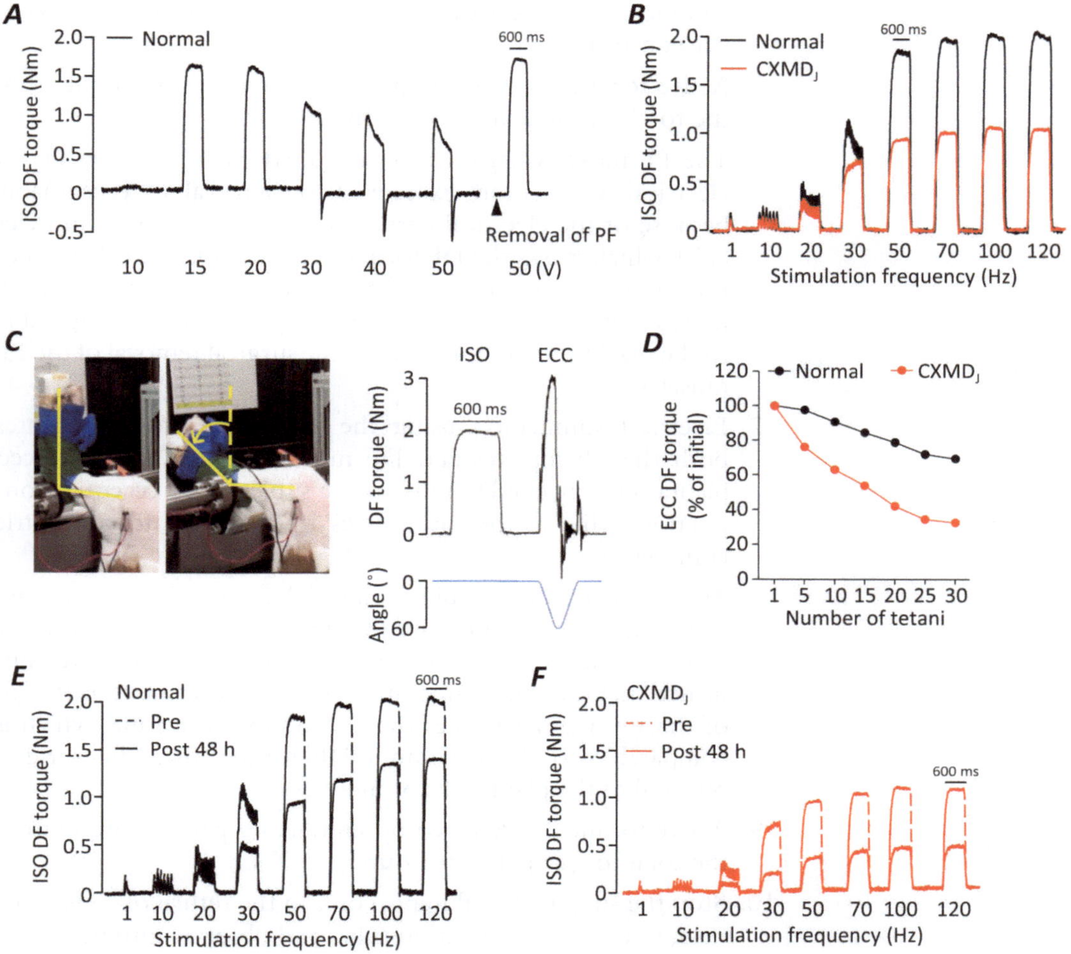

Fig. 6 Typical traces of dorsiflexor torque in a dog. Torque production in isometric contraction (ISO) of dorsiflexor (DF) muscles during increasing stimulation intensities in a normal dog (WT, 42 months old, 10.5 kg body weight) (**a**). Note that PF muscles are recruited above 20 V, which masked the DF torque production. On the other hand, the stimulation intensity just below the recruitment of the PF muscles (i.e., 15 V) evokes the maximum DF torque, which is comparable to the maximum DF torque (i.e., 50 V) after surgical removal of the PF muscles. ISO torque–frequency relationship of DF muscles, with PF muscles intact, in a normal dog (42 months old, black line) and CXMD$_j$ (42 months old, 8.4 kg body weight, red line) (**b**). Comparison between ISO and eccentric contraction (ECC) (0–60° plantar flexion, 290°/s, 50 Hz, 207 ms) in a normal muscle (**c**). Active torque during 30 repeated ECCs, which are applied every 4 s (**d**). Data are expressed as a percentage of the initial value. ISO PF torque before (dotted line) and 48 h (solid line) after 30 ECCs in a normal dog (**e**) and CXMD$_j$ (**f**)

3. ISO torque measurement can be used to set the supramaximal electrical stimulation condition before applying ECCs.
4. Electrical stimulation using the surface electrodes activates both the PF muscles and DF muscles. Thus, during forced dorsiflexion, the PF muscles are subjected to eccentric

contraction, whereas the DF muscles are subjected to concentric contraction.

5. Active torque is calculated by subtracting passive torque from the total torque generated during ECCs.
6. The PF muscles began to be recruited when the intensity of electrical stimulation exceeds a certain value, presumably because of the electrical current spread to antagonist muscles [27], which masks the DF torque (Fig. 4a). On the other hand, the stimulation intensity just below the recruitment of the PF muscles evokes the maximum DF torque, which is comparable to the maximum DF torque after the surgical removal of the PF muscles.
7. Electrical stimulation using the surface electrodes activates both the PF muscles and DF muscles. Thus, during forced plantar flexion, the DF muscles are subjected to eccentric contraction, whereas the PF muscles are subjected to concentric contraction.
8. As described above (Subheading 3.1.3), the PF muscles are activated when the intensity of electrical stimulation exceeds a certain value, which masks the DF torque (Fig. 6a). We confirmed that the stimulation intensity just below the recruitment of the PF muscles evokes the maximum DF torque, which is comparable to the maximum DF torque after the surgical removal of the plantar flexor muscles.
9. Active torque is calculated by subtracting passive torque from the total torque generated during ECCs.
10. Stop the supply of isoflurane, confirm the return of consciousness, remove the intubation tube, and observe until full awakening after measuring the ISO torque following ECCs.

Conflict of Interest

The authors have no conflicts of interest to declare that are relevant to the content of this chapter.

References

1. Jacques MF, Onambele-Pearson GL, Reeves ND, Stebbings GK, Smith J, Morse CI (2018) Relationships between muscle size, strength, and physical activity in adults with muscular dystrophy. J Cachexia Sarcopenia Muscle 9:1042–1052. https://doi.org/10.1002/jcsm.12347
2. Aoki Y, Nakamura A, Yokota T, Saito T, Okazawa H, Nagata T et al (2010) In-frame dystrophin following exon 51-skipping improves muscle pathology and function in the exon 52-deficient mdx mouse. Mol Ther 18:1995–2005. https://doi.org/10.1038/mt.2010.186
3. Minchew EC, Williamson NC, Readyoff AT, McClung JM, Spangenburg EE (2022) Isometric skeletal muscle contractile properties in common strains of male laboratory mice. Front

Physiol 13:937132. https://doi.org/10.3389/fphys.2022.937132

4. Place N, Yamada T, Bruton JD, Westerblad H (2010) Muscle fatigue: from observations in humans to underlying mechanisms studied in intact single muscle fibres. Eur J Appl Physiol 110:1–15
5. Ashida Y, Himori K, Tokuda N, Naito A, Yamauchi N, Takenaka-Ninagawa N et al (2022) Dissociation of SH3 and cysteine-rich domain 3 and junctophilin 1 from dihydropyridine receptor in dystrophin-deficient muscles. Am J Physiol Cell Physiol 323:C885–Cc95. https://doi.org/10.1152/ajpcell.00163.2022
6. Kornegay JN, Bogan DJ, Bogan JR, Childers MK, Cundiff DD, Petroski GF et al (1999) Contraction force generated by tarsal joint flexion and extension in dogs with golden retriever muscular dystrophy. J Neurol Sci 166:115–121. https://doi.org/10.1016/s0022-510x(99)00118-5
7. Cheng AJ, Yamada T, Rassier DE, Andersson DC, Westerblad H, Lanner JT (2016) Reactive oxygen/nitrogen species and contractile function in skeletal muscle during fatigue and recovery. J Physiol 594:5149–5160. https://doi.org/10.1113/jp270650
8. Hollingworth S, Zeiger U, Baylor SM (2008) Comparison of the myoplasmic calcium transient elicited by an action potential in intact fibres of mdx and normal mice. J Physiol 586:5063–5075. https://doi.org/10.1113/jphysiol.2008.160507
9. Fink RH, Stephenson DG, Williams DA (1990) Physiological properties of skinned fibres from normal and dystrophic (Duchenne) human muscle activated by Ca2+ and Sr2+. J Physiol 420:337–353. https://doi.org/10.1113/jphysiol.1990.sp017916
10. Petrof BJ, Shrager JB, Stedman HH, Kelly AM, Sweeney HL (1993) Dystrophin protects the sarcolemma from stresses developed during muscle contraction. Proc Natl Acad Sci USA 90:3710–3714
11. Godfrey C, Muses S, McClorey G, Wells KE, Coursindel T, Terry RL et al (2015) How much dystrophin is enough: the physiological consequences of different levels of dystrophin in the mdx mouse. Hum Mol Genet 24:4225–4237. https://doi.org/10.1093/hmg/ddv155
12. Kiriaev L, Baumann CW, Lindsay A (2023) Eccentric contraction-induced strength loss in dystrophin-deficient muscle: preparations, protocols, and mechanisms. J Gen Physiol 155:e202213208. https://doi.org/10.1085/jgp.202213208
13. Pratt SJP, Shah SB, Ward CW, Inacio MP, Stains JP, Lovering RM (2013) Effects of in vivo injury on the neuromuscular junction in healthy and dystrophic muscles. J Physiol 591:559–570. https://doi.org/10.1113/jphysiol.2012.241679
14. Call JA, Warren GL, Verma M, Lowe DA (2013) Acute failure of action potential conduction in mdx muscle reveals new mechanism of contraction-induced force loss. J Physiol 591:3765–3776. https://doi.org/10.1113/jphysiol.2013.254656
15. Ashida Y, Himori K, Tamai K, Kimura I, Yamada T (1985) Preconditioning contractions prevent prolonged force depression and ca(2+)-dependent proteolysis of STAC3 after damaging eccentric contractions. J Appl Physiol 2021(131):1399–1407. https://doi.org/10.1152/japplphysiol.00463.2021
16. Clarkson PM, Nosaka K, Braun B (1992) Muscle function after exercise-induced muscle damage and rapid adaptation. Med Sci Sports Exerc 24:512–520
17. Blaauw B, Agatea L, Toniolo L, Canato M, Quarta M, Dyar KA et al (1985) Eccentric contractions lead to myofibrillar dysfunction in muscular dystrophy. J Appl Physiol 2010(108):105–111. https://doi.org/10.1152/japplphysiol.00803.2009
18. Araki E, Nakamura K, Nakao K, Kameya S, Kobayashi O, Nonaka I et al (1997) Targeted disruption of exon 52 in the mouse dystrophin gene induced muscle degeneration similar to that observed in Duchenne muscular dystrophy. Biochem Biophys Res Commun 238:492–497. https://doi.org/10.1006/bbrc.1997.7328
19. Shimatsu Y, Katagiri K, Furuta T, Nakura M, Tanioka Y, Yuasa K et al (2003) Canine X-linked muscular dystrophy in Japan (CXMDJ). Exp Anim 52:93–97. https://doi.org/10.1538/expanim.52.93
20. Yuasa K, Nakamura A, Hijikata T, Takeda S (2008) Dystrophin deficiency in canine X-linked muscular dystrophy in Japan (CXMDJ) alters myosin heavy chain expression profiles in the diaphragm more markedly than in the tibialis cranialis muscle. BMC Musculoskelet Disord 9:1. https://doi.org/10.1186/1471-2474-9-1
21. Gregory CM, Vandenborne K, Dudley GA (2001) Metabolic enzymes and phenotypic expression among human locomotor muscles. Muscle Nerve 24:387–393. https://doi.org/10.1002/1097-4598(200103)24:3<387::aid-mus1010>3.0.co;2-m
22. Lindsay A, Southern WM, McCourt PM, Larson AA, Hodges JS, Lowe DA et al (2019)

Variable cytoplasmic actin expression impacts the sensitivity of different dystrophin-deficient mdx skeletal muscles to eccentric contraction. FEBS J 286:2562–2576. https://doi.org/10.1111/febs.14831

23. Yamada T, Kimura I, Ashida Y, Tamai K, Fusagawa H, Tohse N et al (2021) Larger improvements in fatigue resistance and mitochondrial function with high- than with low-intensity contractions during interval training of mouse skeletal muscle. FASEB J 35:e21988. https://doi.org/10.1096/fj.202101204R
24. Kornegay JN, Bogan DJ, Bogan JR, Dow JL, Wang J, Fan Z et al (2016) Dystrophin-deficient dogs with reduced myostatin have unequal muscle growth and greater joint contractures. Skelet Muscle 6:14. https://doi.org/10.1186/s13395-016-0085-7
25. Kornegay JN, Bogan JR, Bogan DJ, Childers MK, Grange RW (2011) Golden retriever muscular dystrophy (GRMD): developing and maintaining a colony and physiological functional measurements. Methods Mol Biol 709:105–123. https://doi.org/10.1007/978-1-61737-982-6_7
26. Garma T, Kobayashi C, Haddad F, Adams GR, Bodell PW, Baldwin KM (1985) Similar acute molecular responses to equivalent volumes of isometric, lengthening, or shortening mode resistance exercise. J Appl Physiol 2007(102):135–143. https://doi.org/10.1152/japplphysiol.00776.2006
27. Neyroud D, Temesi J, Millet GY, Verges S, Maffiuletti NA, Kayser B et al (2015) Comparison of electrical nerve stimulation, electrical muscle stimulation and magnetic nerve stimulation to assess the neuromuscular function of the plantar flexor muscles. Eur J Appl Physiol 115:1429–1439. https://doi.org/10.1007/s00421-015-3124-x

Chapter 7

Characterization and Phenotypic Assessment of Zebrafish Disease Models for Splice Modulation Studies

Holly R. Thomas, Katherine G. English, Matthew S. Alexander, and John M. Parant

Abstract

Zebrafish has emerged as a powerful model in biomedical research as it recapitulates many human disease pathologies. Zebrafish share approximately 70% evolutionarily conserved genes with humans. Recent advances in genome editing have enabled many precision and inducible models that mimic human disorders to be rapidly generated for phenotypic analysis. The large clutch size, ex vivo development, and ability to uptake drug compounds through their skin and gills during early larvae development make the zebrafish an extremely attractive system for high-throughput drug screening. Here, we describe the generation of zebrafish models for splice modulation studies and general phenotypic assessments in human disease modeling.

Key words Zebrafish, Phenotype, CRISPR, Assessments, Functionality, RNA splice modulation

1 Introduction

Zebrafish (*Danio rerio*) are a highly useful animal model for biomedical research laboratories for human disease modeling and therapeutic drug screening and testing. Zebrafish share approximately 70% of known genes with that of humans, and their ex vivo development make them attractive to investigators seeking to perform gain- or loss-of-function genetic studies [1]. Additionally, large offspring sizes (200–300 embryos per mating) and the ability of early stage zebrafish larvae to uptake drug compounds through their skin and gills make the model an attractive organism for high-throughput drug screening [2]. However, it was the discovery that morpholino compounds could be injected into early-stage zebrafish yolk sacs to block or modulate RNA splicing that expanded the use of the zebrafish toward human disease modeling [3]. Guidelines and common techniques have been established to reduce misinterpretations and off-target effects when using morpholinos to study

Yoshitsugu Aoki and Norio Motohashi (eds.), *Splice Switching: Methods and Protocols*, Methods in Molecular Biology, vol. 2963, https://doi.org/10.1007/978-1-0716-4738-7_7,

the splice modulation of RNA in zebrafish [4, 5]. The use of genomic editing tools such as zinc-finger nucleases, TALENs, and CRISPR-Cas9 genomic editing has revolutionized the zebrafish disease modeling field. The rapid adaptation of CRISPR-Cas9 genomic editing has also allowed for the ability to generate fast, precise edits in the zebrafish genome resulting in new models to explore RNA and protein function in a tissue- and temporal-specific manner [6–8]. More recently, there has been some discrepancies between phenotypes reported between zebrafish morphants and homozygous zygotic mutants generated by genomic editing [9]. Thus, it is highly encouraged when evaluating zebrafish phenotypes in morphants to validate any findings in a zebrafish CRISPR- (or ZFN-/TALEN-) generated loss-of-function mutant fish when applicable.

2 Materials

Prepare all genotyping and DNA/RNA analysis reagents under sterile conditions. Ensure proper zebrafish husbandry and euthanasia guidelines are followed depending on the developmental stage and procedure conducted [10]. For zebrafish caudal fin clip genotyping purposes, adult zebrafish should be anesthetized in (approximately 20 mg/L) tricaine (MS-222) before fin clipping and returned into fresh fish water with recirculation promptly. Euthanization of zebrafish should be carried out using a combination of rapid cooling in ice water or a high dosage of tricaine (approximately 200 mg/L or higher) depending on the developmental stage with decapitation via a scalpel blade recommended as a secondary method [10–12].

2.1 Generation of CRISPR-Edited Zebrafish Models for Phenotypic Analysis

- Alt-R™ S.p. Cas9 Nuclease V3 (Cat# 1081058 Integrated DNA Technologies, Coralville, IA).
- Alt-R® CRISPR-Cas9 tracrRNA (Cat# 1072532 Integrated DNA Technologies, Coralville, IA).
- Alt-R® CRISPR-Cas9 crRNA (CRISPR guide RNA) (Integrated DNA Technologies, Coralville, IA).
- Injection needles: Sutter Instruments.
- Thin-walled borosilicate glass w/o filament, outer diameter, 1.0 mm, inner diameter, 0.75 mm (Cat# NC9051667, Thermo-Fisher Scientific).
- Thermo Scientific Samco transfer pipettes (Cat# 13-711-23 Thermo Fisher Scientific).
- Eppendorf Femtotips microloader tips (Cat# 930001007, Thermo Fisher Scientific).
- Agarose (Cat# ICN19398350, Thermo Fisher Scientific).

- Injection mold (Cat# I-34, Adaptive Science Tools, Worcester, MA).
- Petri dishes (Cat# FB0875712 Thermo Fisher Scientific).

2.2 Genotyping of Zebrafish Genomic DNA Using High-Resolution Melting (HRM) Analysis

For adult fish [>60 days post fertilization (dpf)], anesthetize the zebrafish in approximately 20 mg/L mixture of fish water and tricaine (MS-222; pH = 7; Cat# E10521 MilliporeSigma; Burlington, MA, usually kept frozen at −20 °C at a stock of 4 g/L).

- Surgical scalpel (for fin clips) (Cat# 08-920B; FisherScientific; Hampton, NH).
- Genomic DNA digest buffer: 25 mM NaOH (Cat# S5881; MilliporeSigma) and 40 mM Tris-HCl (Cat# 10812846001; MilliporeSigma).
- 96-well PCR plates for fin clip genomic DNA extraction (Cat# 14-230-234; ThermoFisher Scientific).
- 10× LC Green Plus Melting Dye for high-resolution melting analysis (Cat# BCHMASY-0005; BioFire Defense LLC, Murray UT).
- 0.05 μL GenScript Taq, 0.2 μL 10 mM dNTP Mix, 1 μL 10× reaction buffer, and 0.3 μL 15 mM $MgCl_2$ (Cat# E00101, Genscript; Piscataway, NJ).
- 0.3 μL of each oligo/primer (10 μM) (Integrated DNA Technologies; Coralville, IA, or Life Technologies, Carlsbad, CA).
- 1.5 μL zebrafish genomic DNA from a fin biopsy.
- Nuclease-free water (Cat# AM9939; Invitrogen; Waltham, MA).
- Mineral oil (Cat# O121-1, ThermoFisher Scientific).
- Skirted hard shell 96-well PCR plate for PCR reaction and HRM analysis (Cat# **HSP9665, Bio-Rad; Hercules, CA).**
- Sealing tape for optical assays (Cat# 2239444, Bio-Rad; Hercules, CA).
- Centrifuge to spin down 96-well plate.
- PCR machine: BioRad C1000 Touch Thermal Cycler or BioRad CFX Opus 96.
- LightScanner System instrument (Idaho Technology; Salt Lake City; UT) or BioRad CFX Opus 96.
- Recovery isolation tanks for genotyping (Cat# DC-96; R&D Aquatics; Columbia, MD).

2.3 DanioVision Locomotor Assessments

- DanioVision observation chamber with EthoVision XT software (Noldus; Leesburg, VA).
- 48-well flat-bottom Whatman polypropylene clear plate (Cat# 7701-5500; Tisch Scientific) for optical imaging.

- Universal microplate lid, unilid, clear polystyrene (Cat# 7704-1001; GE Healthcare).

2.4 Touch Evoke Response

- Pronase solution in E3 medium. Pronase (Cat# 10165921001; Roche Applied Science, Penzberg, Germany) made to 5 mg/mL stock and diluted to 1 mg/mL.
- To make E3 medium (alternative versions exist), use 5 mM NaCl (S7653; Millipore Sigma); 0.17 mM KCl (Cat# BP366–1; FisherScientific); 0.33 mM CaCl2 (Cat# 4901; Millipore Sigma); 0.33 mM $MgSO_4$ (Cat# M7506; Millipore Sigma); and 0.00001% (w/v) methylene blue solution (Cat# 319112; Millipore Sigma).
- Nikon SMZ1500 stereomicroscope with an attached camera (CMOS C-Mount Microscope Camera; United Scope LLC; Irvine, CA).
- Probe stimulus (gel loading pipette tips work well) (Cat# 07-200-288; Fisher Scientific).
- Pasteur pipet (Cat# 22-378893; Fisher Scientific).
- Dumont forceps #5 (Cat# 11251–10; Fine Science Tools; Foster City, CA).
- 10 cm Corning TC-treated culture dishes (Cat# 08-772-22; Fisher Scientific).

2.5 Skeletal Muscle Birefringence

- Transfer pipete (Cat# S304673; FisherScientific).
- Nikon SMZ1500 stereomicroscope (or an equivalent microscope) with a polarizing filter (Nikon; Tokyo, Japan).
- Detachable Nikon circular polarizer II filter (52 mm) (Nikon).

2.6 RNA-Quantification and RNA Splicing Evaluation

- TRIzol reagent (Cat# 15596026; ThermoFisher Scientific).
- Chloroform (Cat# J67241-AP; ThermoFisher Scientific).
- DNase I (Cat# 89836; ThermoFisher Scientific).
- SuperScript III first strand synthesis system (Cat# 18080051; ThermoFisher Scientific).
- Lysing matrix D 2 mL tube (Cat# 116913050-CF; MP Biomedicals LLC; Santa Ana, CA).
- Benchmark BeadBlaster 24 microtube homogenizer, 112 V; TE Equipment, Long Branch, NJ).
- High-Capacity cDNA reverse transcription kit (Cat# 4368814, Life Technologies).
- Takara Ex Taq DNA polymerase (Cat #50-443-968, ThermoFisher Scientific).
- 0.3 μL of each oligo/primer (10 μM) (Integrated DNA Technologies; Coralville, IA, or Life Technologies, Carlsbad, CA).

- 0.2 mL PCR tubes (Cat# 14-230-225, ThermoFisher Scientific).
- Agarose Cat# ICN19398350, Thermo Fisher Scientific).
- Gel electrophoresis apparatus.
- Promega Wizard SV gel and PCR cleanup system (Cat# A9282, ThermoFisher Scientific).
- NanoDrop ND-1000 UV–Vis spectrophotometer (ThermoFisher Scientific).
- QuantStudio 5 real-time PCR thermocycler (ThermoFisher Scientific).
- PowerTrack SYBR Green Master Mix (2×) (Cat# A46012; ThermoFisher Scientific).

3 Methods

3.1 CRISPR Guide Design: The First Step in Genome Editing Is to Design the Optimal Guide Targeting the Desired Site

1. First, determine the type of model you want to design. For disease models that require a gene knockout (KO), it is preferred to target an exon 5′ in the gene but after the ATG, such that if a protein is produced it will be very small. For splice site modeling, the guide can target across an exon–intron junction. The consensus AGgt of the splice donor and agG of the splice acceptor often serve as good PAM sites (see example in Fig. 1a).
2. There are a number of guide design tools, but we primarily use the IDT custom design web site (https://www.idtdna.com/site/order/designtool/index/CRISPR_CUSTOM).
3. Click zebrafish in the drop-down menu, then paste the nucleotide region of interest, and click design. If designed to splice acceptor, include intronic sequence at the exon junction.
4. In general, it is best to choose the guide with the best on-target and off-target score. An example of the results for the guide design against *cenpj* is in Fig. 1b. Guides with an on-target score of >50 often work well. However, choosing a guide based on the on-target and off-target scores can be misleading, in that we have used guides with on-target scores less than 30 that showed efficient targeting. If using a guide RNA with a low off-target score, this is usually a minor concern since you can breed out any off-target mutations.

3.2 Generation of CRISPR-Edited Zebrafish Models for Phenotypic Analysis

1. Prepare an RNP mixture: A 3 μM gRNA solution is obtained through diluting 3 μL of 100 μM crRNA and 3 μL of 100 μM tracrRNA into 94 μL of nuclease-free duplex buffer, heating at 95 °C for 5 min, and then cooling to room temperature. 0.5 μL of Cas9 protein is diluted with Cas9 working buffer (20 mM

A.

Atm x8	PAM sgRNA
Wild-type	taattgcatttctttcccagCTGTTCAATCAGAGCACTCGTGTTCAGGAAGTGACATCATCTGT
3bp deletion (Δ3)	taattgcatttctttcccca----GTTCAATCAGAGCACTCGTGTTCAGGAAGTGACATCATCTGT
5bp deletion (Δ5)	taattgcatttctttc-------CTGTTCAATCAGAGCACTCGTGTTCAGGAAGTGACATCATCTGT
Δ3 + 5 bp	taattgcatttctttccccag------AGCACTTCAATCAGAGCACTCGTGTTCAGGAAGTGACATCATCTGT

AldoB x3	sgRNA PAM
Wild-type	GGACAAGGGCATCGTAGTTGGTATCAAGgtgagattgcagtgcatctgtgttcatgcat
4bp deletion (Δ4)	GGACAAGGGCATCGTAGTTGGT-------------gagattgcagtgcatctgtgttcatgcat
8bp deletion (Δ8)	GGACAAGGGCATCGTAGTTGGTA--------Ggtgagattgcagtgcatctgtgttcatgcat
Insertion of 1bp (+1)	GGACAAGGGCATCGTAGTTGGTAGTCAAGgtgagattgcagtgcatctgtgttcatgcat
Insertion of 17bp (+17)	GGACAAGGGCATCGTAGTTGGTAGGTGAGATTGCAGTGCATCAAGgtgagattgcagtgcatctgtgttcatgc

Cenpj x1	sgRNA PAM
Wild-type	TAAATACCCGGCTCGACAACCCAAACTAAGgtagggcatatttactcacaactctgcagcga
5bp deletion (Δ5)	TAAATACCCGGCTCGACAACCCAAACTAAGg--------catatttactcacaactctgcagcga
5bp deletion (Δ5)	TAAATACCCGGCTCGACAACCCAAACTA---------gggcatatttactcacaactctgcagcga
9bp deletion (Δ9)	TAAATACCCGGCTCGACAACCCAAACT--------------catatttactcacaactctgcagcga

Usp16 x4	sgRNA PAM
Wild-type	TCTCAGCCTGGACGTCTGGAGTGTCTGGtcagtcacacacacacacagatcctcagtgaag
9bp deletion (Δ9)	TCTCAGCCTGGA----------------TGTCTGGtcagtcacacacacacacagatcctcagtgaag
11bp deletion (Δ11)	TCTCAGCCTGGACGTCTGGAGT----------------cacacacacacacagatcctcagtgaag

B.

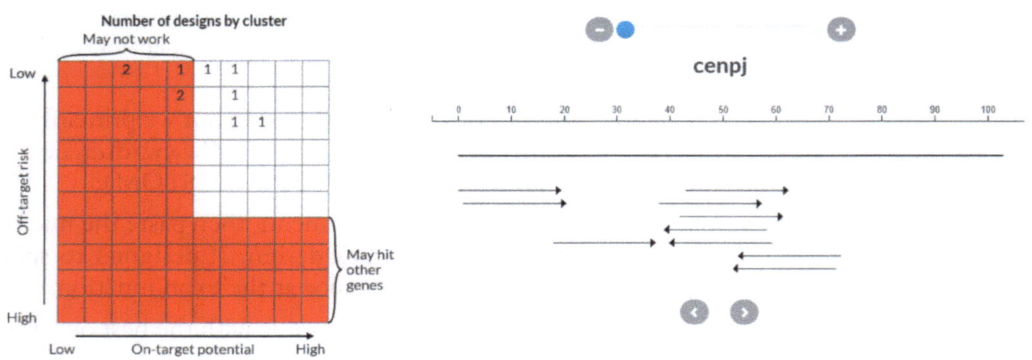

Position	Strand	Sequence	PAM	On-target score	Off-target score
43	+	CGACAACCCAAACTAAGGTA	GGG	65	93

Fig. 1 CRISPR guide examples located at the exon/intron or intron/exon junctions to target splice sites, the alleles detected using each CRISPR guide, and an example of IDT guide design results. (**a**) The guide RNA (underlined) and PAM sequence (highlighted) are indicated; exon sequence is shown as uppercase letters, and intron sequence is shown as lowercase letters. "-" represents deleted nt, while red fonts are inserted nt. (**b**) Example of typical output from the IDT CRISPR design tool, indicating the location of the guides relative to the input sequence and the on-target and off-target scores. The CRISPR guides in the white boxes would be more ideal to try for targeting, whereas those guides located in the red-shaded boxes have a low on-target score or a high off-target score

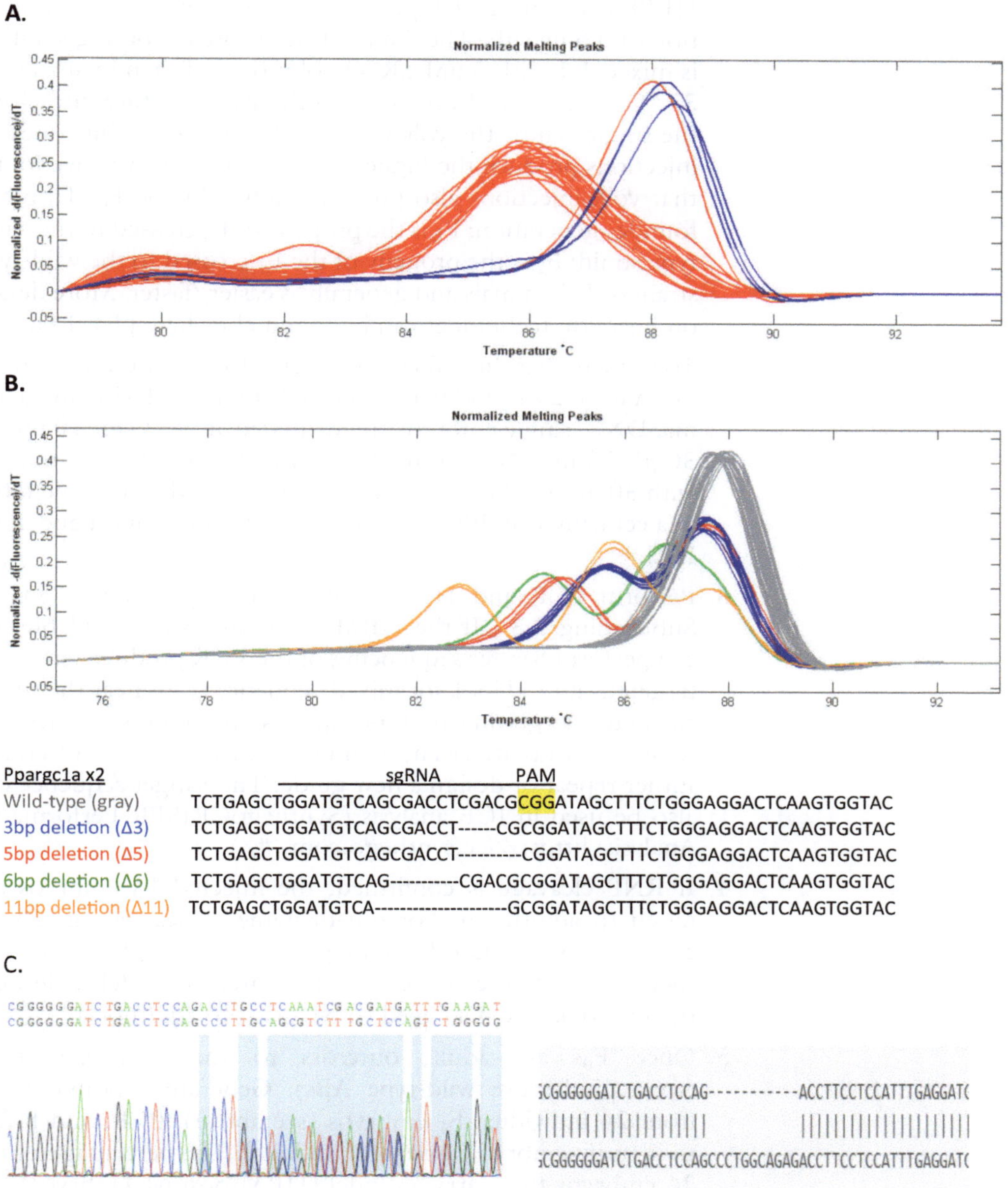

Fig. 2 High-resolution melt (HRM) curve analysis to determine CRISPR guide effectiveness, genotype, and allele identification. (**a**) HRM curves of injected F_0 embryos indicate that CRISPR effectively cuts at the desired site. Wild-type uninjected embryos are indicated by blue curves, and injected F_0 embryos (targeted with CRISPR) are indicated by red curves. The one wild-type-looking red curve represents poorly injected embryos or one that lacks cleavage. (**b**) HRM curves of F1 (or beyond—nonchimeric) can be used to genotype fish and identify different alleles. Wild-type curves are gray, and the different alleles are indicated by blue, red, green, and orange curves. (**c**) Example of Poly Peak Parser results. The ABI sequence chromatogram file and the reference sequence can be inserted into the website, and the results generated will indicate the exact sequence of heterozygous mutation

HEPES; 150 mM KCl, pH 7.5) to yield a working concentration of 0.5 μg/μL. The diluted Cas9 protein working solution is mixed 1:1 with 3 μM gRNA solution and then incubated at 37 °C for 10 min. Inject 1 nL of the RNP mixture into either the single cell or the yolk of one-cell embryos. While one-cell injections produce the highest indel frequency, we have found that yolk injections also produce high-chimeric F_0 (F_0 is the founder generation; F_1 is the progeny of F_0 crossed to the wild-type strain; F_2 is the progeny of the F_1 crossed to the wild-type strain or F_1) animals and generally is easier/faster. More details on injection techniques can be found elsewhere [13, 14].

2. To validate that the guide RNA is effective, we isolate genomic DNA from 24 F_0 individual injected embryos. To isolate genomic DNA, single embryos are incubated at 98 °C for 10 min in 30 μL 25 mM NaOH in a 96-well plate and then neutralized with 30 μL of 40 mM Tris–HCl. Spin down the genomic DNA in a centrifuge at 2000 rpm for 1 min to mix NaOH and Tris–HCl.

3. Perform PCR and HRM analyses (example in Fig. 2a; *see* Subheading 3.3). If the HRM technique is not available, we can perform Sanger sequencing of the PCR product across the targeted site and look for mixed-sequence reads near the site of cleavage, suggesting multiple species are being sequenced. If sequence reads are clean, then guide injections did not cleave. Either repeat or design a new guide. The Sanger sequence can also be used in ICE analysis (SynthegoCRISPR Performance Analysis; https://ice.synthego.com/#/).

4. If RNP cleavage is confirmed, the injected F_0 embryos are raised to adulthood. For out-of-frame alleles, raising 20–30 embryos to adulthood is enough to get multiple indels. If a specific size, frame of, or specific content of indel is desired, then more larvae may need to be raised to adulthood.

5. Once F_0s are adult, outcross to the wild-type strain (we typically use wild-type ABs). Generate genomic DNA from 24 individual F_1 embryos (see above), PCR, and HRM to identify embryo carriers for an indel. The percentage of the 24 embryos that carry an indel (HRM-positive embryos) can be used to predict the germline transmission rate of that F_0. In addition, different HRM curves often represent different indels (Fig. 2b). We often choose to raise progeny (50 embryos per F_0 and raise progeny from 3–5 F_0) from F_0 with the highest germline rate or F_0 offspring that produce many alleles.

6. Once F_1 embryos are raised to adulthood, we isolate genomic DNA from F_1 tail clippings. First, anesthetize the fish in approximately 20 mg/L mixture of fish water and tricaine for 30 s. Watch for the gills of the fish to stop moving or for the

fish to float on their back with their belly up. Each fish is placed in a sterile padding, and using a sterile scalpel, a small fin biopsy is performed (usually no more than 8 mm in diameter). Place the fin clip biopsy in the 96-well plate with sterile forceps. After biopsy, the anesthetized fish is placed in a recovery tank of fish water, before being placed in an isolation tank with recirculating water. To isolate genomic DNA from the tail biopsy, add 40 μL of 25 mM NaOH to the 96-well plate; heat to 95 °C for 20 min and then neutralize with 40 μL of 40 mM Tris–HCl in each well. The 96-well plate with genomic DNA can be stored at 4 °C (*see* **Note 1**).

7. Next, we perform genomic DNA PCR and HRM analysis (see below) to identify the carriers of an indel. For those F_1 that are positive, a secondary PCR (300–500 base pairs/bp) product across the cleavage site is Sanger-sequenced. From this Sanger sequence, we use the Poly Peak Parser website (http://yosttools.genetics.utah.edu/PolyPeakParser/) to identify the exact allele sequence (Fig. 2c). To use this website, only the expected wild-type sequence from ensemble.org and the Sanger Sequencing ABI file are needed, and the software will display the heterozygous allele sequence for that particular sample [15]. To select the experimental allele to use, we recommend maintaining two lines with frame shifts in the +1 and + 2 frames. Larger indels are often preferred since they allow for more genotyping options later on (i.e., allele-specific PCR). If interested in potential hypomorphs/weak alleles, we maintain the +3 frame as well, with larger in-frame alterations likely being more deleterious.
8. To avoid CRISPR derived off-target mutations in the background, we always breed the identified individual F_1 to the wild-type strain one more time to derive a F_2 heterozygous population, in which each fish have approximately 75% "clean" genome that has never been exposed to CRISPR reagents.

3.3 Genotyping of Zebrafish DNA Using High-Resolution Melting (HRM) Analysis

1. Two applications of HRM: (1) detection of heterozygous mutations and (2) genotyping homozygous wild-type, heterozygous, and homozygous mutant animals.
2. Designing HRM primers. The detection sensitivity of HRM is best with smaller amplicons, particularly when differentiating homozygous wild-type from homozygous mutants. However, smaller amplicons may also bias against the detection of larger indels when used for mutation detection. We find that approximately 110 bp amplicons are best for both attributes, with the cleavage site ideally in the middle of the amplicon. We use Primer3 (https://primer3.ut.ee/) to design primers. The predicted wild-type sequence is obtained and entered from ensembl.org, and brackets ([and]) are placed in sequence to

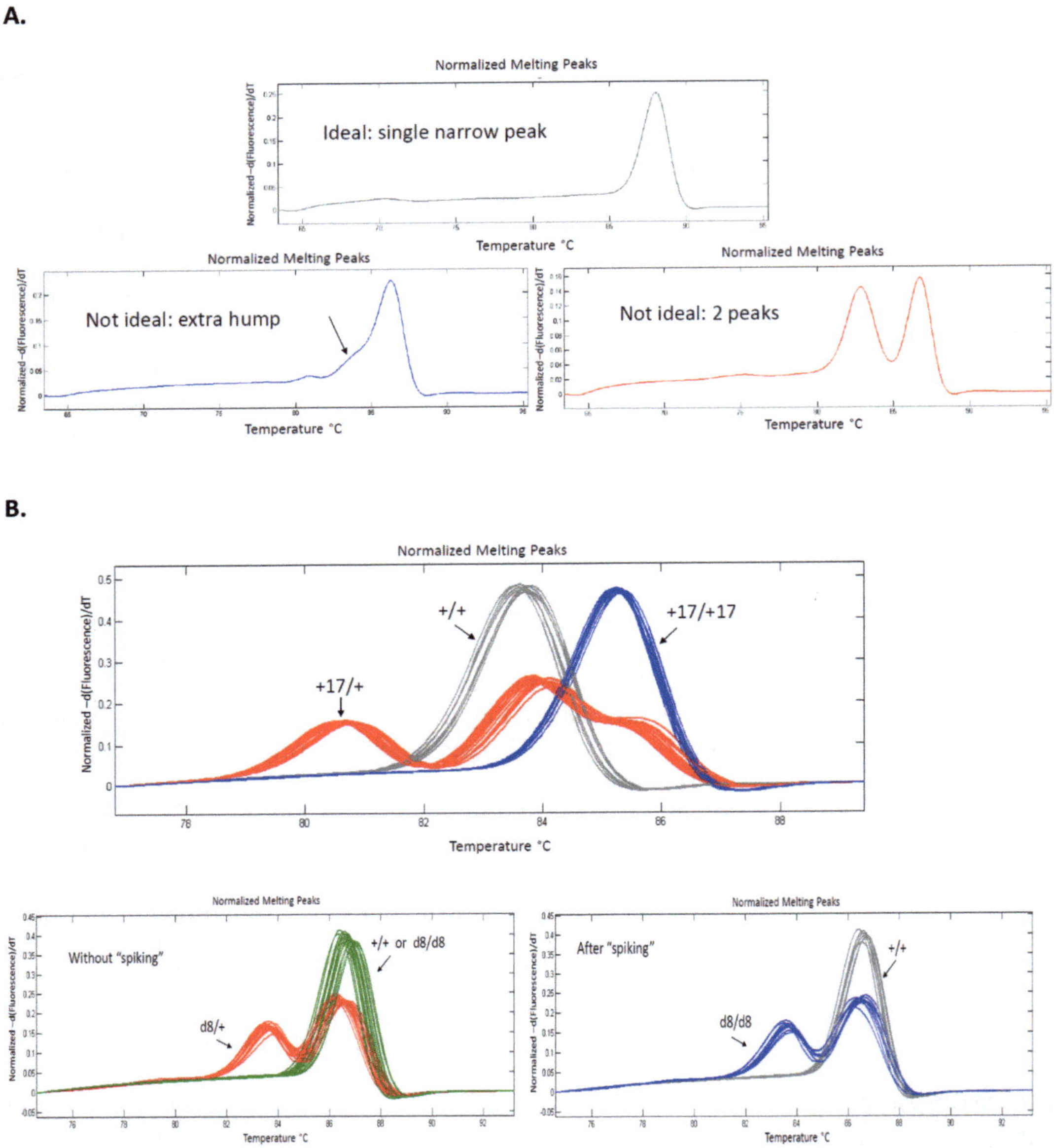

Fig. 3 Representative HRM melting curve peak analysis for CRISPR allele genotyping. (**a**) HRM peak analysis of representative HRM melt curves: ideal (a single, narrow peak), not ideal (two or more peaks), and not ideal (extra, smaller hump on a large, single peak). Normalized fluorescence/dT plotted against the melting curve temperature (°C) is shown. (**b**) Example of an identified mutation in which the homozygous wild-type (gray), heterozygous (red), and homozygous mutant (blue) melting curves easily separate at different melting temperatures (top panel). Example of an identified mutation in which the homozygous wild-type and homozygous mutant melting curves have a similar melting temperature and cannot be easily distinguished (green melting curves—left bottom panel). After spiking with wt DNA, the homozygous wild-type (green) melting curves appear as a single peak, whereas the homozygous mutant melting curves now appear heterozygous for the identified mutation (orange curves—right bottom panel)

indicate what specific region must be included in our desired amplicon. The length of the desired amplicon can be determined, along with the primer length and primer Tm, among many other options, but we typically aim for primers with an approximate Tm of 60 °C (Fig. 2).

3. Testing the primer set for HRM. It is useful to first test the HRM on genomic DNA from (on average) 24 wild-type embryos to determine if the HRM analysis is favorable (results in a narrow single peak). There are a couple of confounding effects that could occur: (1) if there are additional products amplified, this will result in extra HRM curves and (2) if there are SNPs in this region of some wild-type embryos, we will have heterozygous curves that can be false indel positives in future analysis. Examples of good and poor HRM curves are displayed in Fig. 3a.
4. PCR is performed with HRM primers on genomic DNA in a white-welled skirted 96-well plate. Each PCR reaction contains 1 μL of genomic DNA and 9 μL of a Master Mix and 30 μL of mineral oil. The PCR master mixture includes: 1 μL of LC Green Plus Melting Dye (Biofire Defense, BCHM-ASY-0005), 1 μL of 10× enzyme buffer, 0.2 μL of dNTP mixture (10 mM each), 0.3 μL of $MgCl_2$, 0.3 μL of each primer (10 μM), 1 μL of genomic DNA, 0.05 μL of GenScript Taq (E00101), and water up to 9 μL.
5. The PCR plate is centrifuged at 2000 rpm for 1 min and then placed in the PCR machine. The typical PCR reaction protocol is 98 °C for 30 s, then 45 cycles of 98 °C for 10 s, 59 °C for 20 s, and 72 °C for 15 s, followed by 95 °C for 30 s, and then rapid cooling to 4 °C. The important part of HRM is that in place of a long 72 °C extension, at the end of the cycles, a 95 °C denaturation and a rapid cooling to 4 °C step is added to promote the formation of heteroduplexes.
6. For the Idaho technology light scanner HRM machine, we transfer the PCR plate to the machine and perform the HRM scan. Melting curves are generated and analyzed using the LightScanner instrument over a 65–95 °C range. As an alternative strategy, a BioRad CFX Opus 96 system (*see* **Note 2**) can be used with the same PCR as described above, but after the final 95 °C denaturation step, we allow the plate to equilibrate at 65 °C for 1 min and then perform the melt curve analysis by slowly heating the plate over a 65–95 °C range and allowing the machine to read fluorescence at every 0.5 °C interval. While identifying heterozygous genotypes is relatively consistent due to the heteroduplex, differentiating homozygous wild-type and homozygous mutants can be difficult if there is not much T_M difference. For these instances, perform a second PCR and use

1 μL of genomic DNA and 1 μL of wild-type "spike" DNA such that homozygous wild-type samples will be homozygous by HRM, while homozygous mutant samples will now appear heterozygous by HRM. Examples of "spike" results are shown in Fig. 3b.

3.4 RNA Quantification, RNA Splicing Evaluation, and RNA NMD Analysis

1. Euthanize zebrafish larvae using high-dose tricaine (approximately 200 mg/L or higher) and ice water cooling methods [11].
2. Total RNA can be extracted from euthanized zebrafish larvae or tail biopsy by placing the specimen into bead matrixes using mechanical homogenization along with TRIzol reagent following standard total RNA isolation protocols [23].
3. Quantify the amount of total RNA in a spectrophotometer (*see* **Note 3**).
4. Use High-Capacity cDNA Reverse Transcription Kit (Life Technologies, 4368814) to generate cDNA (consult manufacturer's protocol). Mix 50–100 ng of total RNA in water up to 10 μL with 2 μL of 10× RT buffer, 0.8 μL of 25× dNTP mix, 2 μL of 10× random primers, 1 μL of reverse transcriptase, and 4.2 μL nuclease-free water. Incubate at 25 °C for 10 min, 37 °C for 2 h, 85 °C for 5 min, and then store at 4 °C.
5. To determine the changes in mRNA levels, we perform SYBR-Green-based qPCR analysis with the dissociation curve melt analysis to ensure only singular product amplification and the lack of primer dimer (Fig. 4a). Standard melt curve analysis can be performed using the manufacturer's protocol (e.g., ThermoFisher Scientific) or following the reference guidelines [16]. SYBR Reference sets of zebrafish housekeeping transcript probes should be employed to normalize transcripts and expression levels [24–26]. Transcripts such as β-actin-2 (*actb2*) and eukaryotic translation elongation factor 1 α-1 (*eef1a1*) usually work well and display no variability in the expression level for most expression reference controls.
6. For analyzing spice variants, PCR primers are designed outside of the exon of interest, ideally two exons in front and two exons downstream that produce a PCR product between 300 and 500 bp. The PCR product is purified (using the Promega Wizard SV Gel and PCR Cleanup System) and submitted to amplicon next-generation sequencing (Azenta (https://www.genewiz.com/en/Public/Services/Next-Generation-Sequencing) would be one example of this service). From the next-generation sequencing (NGS) data, you can access how often alternative splice variants occur and the exact splice site. We often use the CRISPR RGen Tools Cas Analyzer website (http://www.rgenome.net/cas-analyzer/#!) to analyze the

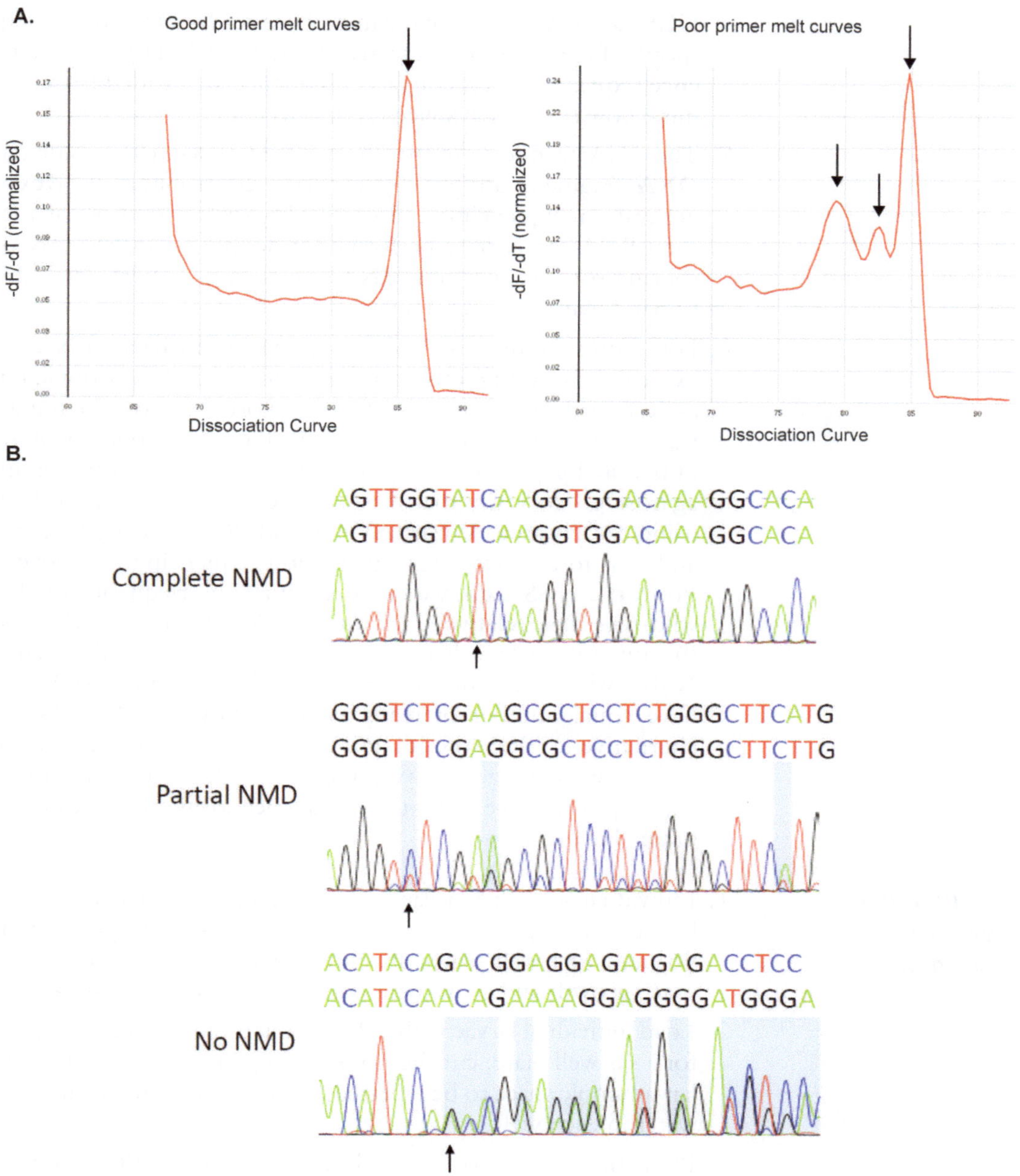

Fig. 4 RNA splicing and nonsense-mediated decay (NMD) DNA chromatogram analysis. (**a**) SYBR-green-based qPCR analysis with dissociation curve melt analysis to ensure only singular product amplification and the lack of primer dimer. (**b**) Sequence chromatograms compared to the area under the curves and indicating if there is complete NMD (top), partial NMD (middle), or no NMD (bottom). Arrows indicate the position on the chromatogram of the start of the known heterozygous mutation

NGS data and determine the number of reads obtained for each specific PCR amplicon detected. Only the FASTQ files, reference sequence, and guide target sequence are needed to determine how often each splice variant occurs.

7. The RNA products can then be resolved on agarose gels or via DNA fragment analysis by capillary electrophoresis. Next-generation sequencing and cDNA library preparation of the total RNA can also be performed if desired to analyze RNA-splicing patterns, perform RNA-sequencing, or perform additional downstream RNA analyses.
8. For nonsense-mediated decay (NMD) analysis, RNA is extracted from a heterozygous tail biopsy. cDNA is obtained as described above, and RT-PCR is performed across the mutation site, with primers that produce an approximately 300–500 bp product [17]. The PCR product is run on an agarose gel to verify that it is a single PCR product, purified using the Promega Wizard SV Gel and PCR Cleanup System, and sent to amplicon next-generation sequencing (as above). From the NGS data, we can determine the frequency of the wild-type allele relative to the mutant allele. Complete NMD of the mutant allele will reveal no mutant reads, while partial NMD will have reduced mutant reads, and no NMD will have equal number of reads as the wild-type DNA. As an alternative to NGS, the RT-PCR product can be submitted to Sanger sequencing, and we can compare the amplitude of the wild-type sequence and mutant sequence within the chromatogram (Fig. 4b).

3.5 DanioVision™ Locomotor Assessments

1. Individual 4–7 dpf larvae are placed in 48-well plates. The larvae can be placed in 6-, 12-, 24-, and 48-well plates, but 96- and 384-well plates are not preferable due to the larval space growth and movement constraints [13] (*see* **Note 4**).
2. Load individual larvae with a P1000 micropipette set to 300 μL for a 48-well plate; cut the tip of the pipette to allow zebrafish larvae ample space to be pipetted. Maintain the same volume of water in each well.
3. Place the plate into DanioVision™ (version DVOC-0040), and close the inner and the outer lids (Fig. 5a).
4. Adjust the settings on the EthoVision (version 16) software to select the 48-well plate, live tracking, and the desired time for the run. Be certain to make sure that the arena settings are correct for each 48-well plate to ensure that the outline of each arena is correctly aligned.
5. For a standard zebrafish larva locomotor analysis, use User Set 2 with a defined camera resolution at 640 × 480 for a 12 mm lens at 60 frames/second. For additional adjustments, seek

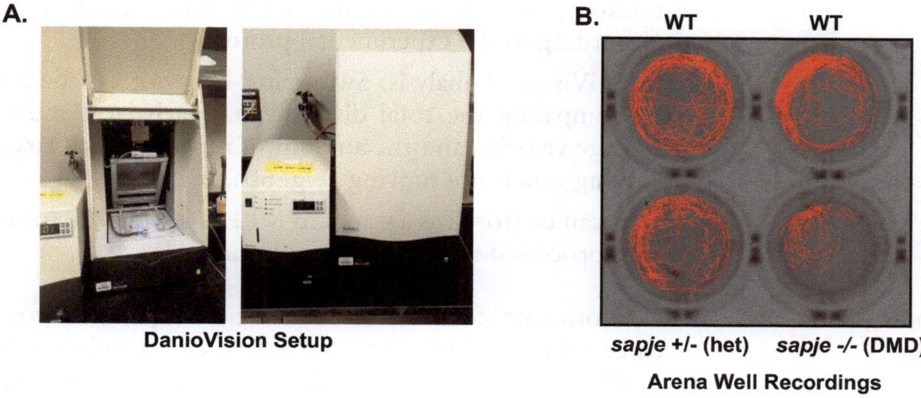

Fig. 5 Assessment of zebrafish locomotor phenotypes using DanioVision activity tracking. (**a**) DanioVision standard setup used for locomotor assessments. Plates are kept at 28 °C within the humidified chamber during the assessment. Open and closed box coverings for activity tracking. (**b**) Activity tracings of wild-type (*AB*), *sapje* +/− (heterozygote), and *sapje* −/− (homozygote) DMD mutant zebrafish. DanioVision recordings were performed at 5 dpf

guidance from the user manual and purchased system camera setup.

6. Under the setting "Validate Setup", make sure the arena settings are validated by outlining each well with no overlap and properly calibrating the plate size. Individual wells can be highlighted, and wells without larvae can be excluded. Individual areas can be highlighted under the tabs "Zone Groups" and can be excluded in the later analysis in EthoVision.
7. Run the standard trial control settings. White light stimulus can be used for 10 min after the zebrafish are loaded in the device for habituation, listed under "Default" or "White Light Routine".
8. Imaging should be performed over a fixed time interval (15 min); allow for re-run with additional time (15 min) for larva adjustments.
9. Alternating light routines can be performed to measure light–dark responses. Under "Trial Control Settings", select "New" and adjust "Settings" under "Start-Stop Trial Control Rule" to create a new protocol. The user can add alternating light–dark periods (example: 15 min light, 10 min dark, 10 min dark, 15 min light). Light intensity can also be adjusted under "Hardware Action" window under the "Intensity level (from 0% to 100%; from 0 to 5450 lux).
10. To acquire data, we should make sure that we have planned the appropriate time intervals under the Trial list and highlighted the appropriate arenas (or wells in which the larvae are to be

tracked), as well as set the appropriate detection settings (according to the experimental protocol).

11. DanioVision™ analysis: Swimming behavior can be analyzed by comparing the total distance that each larva moved, the average velocity, and the amount of time that each larva spent moving versus not moving (Fig. 5b).
12. Files can be stored as a standard MPEG-4 video file format and later processed in EthoVision software.

3.6 Touch Evoke Response

1. Dechorionate >24 h post fertilization (hpf) embryos. Place embryos with chorion in a 1.5 mL microcentrifuge tube with 1 mL of E3 medium and add 30 μL of 1 mg/mL pronase. Allow to incubate at room temperature for 10 min. With a Pasteur pipet, gently pipet the embryos up and down to tear open the chorions. Wash the embryos by removing the pronase solution and adding 1 mL of E3 media; gently pipet up and down and remove E3 media. Add fresh E3 media to wash the embryos for a total of three washes and transfer the embryos to a clean 10 cm Petri dish containing E3 media. If necessary, using the Dumont forceps, gently remove the chorions. Handle dechorionated zebrafish larvae with care using the Pasteur pipets as they are susceptible to damage.
2. Place the dechorionated larvae at the center of the field of the plate with camera recording. Make sure that the camera settings record at a frame rate of greater than 30 Hz with the goal of evaluating image captures at 15 ms post stimulus with the mechanosensory probe. All wild-type fish should move away from the stimuli after 300 ms, and the experiment should be repeated at least five times per fish if possible for reproducibility.
3. Time-lapsed still images can be processed on commercial software (e.g., Nikon NIS Elements) or freely available ImageJ software [14]. Zebrafish muscle mutants such as the *sapje/sapje-like* (*dmd*), dystroglycan (*dag1*), and laminin alpha-2 (*lama2*) will likely display prominent touch-evoke escape defects [15–17].
4. Swimming behavior can be quantified by comparing the average distance the larvae swim during the fixed time (e.g., 300 ms) between the start (0 ms) and end time (300 ms).
   (a) Repetitive stimulus on the same larvae (e.g., greater than five stimulations) should be avoided to both avoid damage to the larvae and habituation.

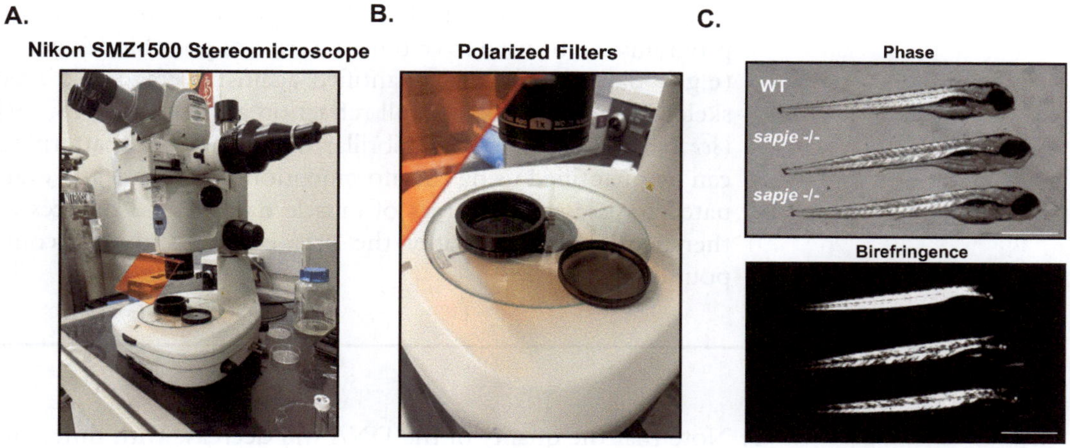

Fig. 6 Standard microscopy setup for zebrafish assessment of skeletal muscle birefringence. (**a**) Nikon SMZ1500 stereoemicroscope used for zebrafish skeletal muscle birefringence. (**b**) Polarized lenses, with the larger lens attached to the upper objective lens, while the bottom (thinner) lens has the tricaine-anesthetized zebrafish larvae. (**c**) Representative phase and birefringence microscopy images of one WT (*AB*) and two *sapje* −/− (homozygote) mutant zebrafish taken at 4 days post fertilization (dpf). Scale bar demarcates 500 μm

3.7 Skeletal Muscle Birefringence

Skeletal muscle larvae should be evaluated between 4 and 5 dpf as several zebrafish skeletal muscle mutants show muscle birefringence phenotypes during these developmental time points [18].

1. Place the tricaine-anesthetized larvae on a polarized filter using the transfer pipet with minimal fish water (Fig. 6).
2. Adjust the anesthetized larvae laterally to allow for maximal polarized light to refract off the zebrafish dorsal skeletal muscle. Keep the larvae as flat as possible against the lens. If the larva has a curvature, try to hold a probe against it to keep it flat when imaging.
3. Rotate the top lens, with the embryo of interest in view, until the axes of polarization of the two lenses are oriented at 90° from one another and the background is completely dark when imaging.
4. Image the larvae using standard bright-field imaging. Multiple images of the same fish (at least $n \geq 3$ if possible) and of large numbers of larvae are obtained per genotype cohort or condition when possible.
5. Analysis of birefringence. Wild-type zebrafish will have the maximal birefringence when imaging, and pictures acquired can be used as reference for quantification in ImageJ or other software programs. Zebrafish with detached myofibers from sarcolemma membranes will have poor skeletal muscle birefringence patterns (e.g., dystrophic zebrafish models; Fig. 6c), and

this can be quantified [18, 19]. ImageJ and other software programs can be used to outline the areas of poor birefringence (e.g., "dark spots") and quantified against average wild-type skeletal muscle regions of full refraction as described [18, 19] (*see* **Note 5**). Unique myofibrillar disorganization patterning can be quantified using the aforementioned software and evaluated to assess the degrees of muscle damage and degrees of therapeutic correction after the larva exposure to drug compounds [21, 22].

4 Notes

1. Note that the quality of the DNA will decrease with time, but we have been able to PCR amplify this genomic DNA at least 3 months after storage at 4 °C.
2. For BioRad CFX Opus 96, the PCR and HRM scan can be performed using the same machine.
3. 260/280 Nanodrop spectrophotometer measurement greater than 2.0 is usually considered pure, although 1.8 or higher can also be used for analysis.
4. Outer edges (especially corners) of the plate are generally not preferred due to "edge effects" in which the detector might miss or not accurately record the larvae.
5. The majority of these procedures are for early-stage zebrafish larvae but can be adapted for adult zebrafish, excluding skeletal muscle birefringence and other assays that are optimized for early developmental time points. Some skeletal muscle congenital myopathy zebrafish such as myotubularin morphants (*mtm1* and *mtmr12*) have severe birefringence patterns compared to dystrophic models [20].

Conflict of Interest

The authors have no conflicts of interest to declare that are relevant to the content of this chapter.

References

1. Howe K et al (2013) The zebrafish reference genome sequence and its relationship to the human genome. Nature 496(7446):498–503
2. Patton EE, Zon LI, Langenau DM (2021) Zebrafish disease models in drug discovery: from preclinical modelling to clinical trials. Nat Rev Drug Discov 20:611
3. Nasevicius A, Ekker SC (2000) Effective targeted gene 'knockdown' in zebrafish. Nat Genet 26(2):216–220
4. Bill BR et al (2009) A primer for morpholino use in zebrafish. Zebrafish 6(1):69–77

5. Stainier DYR et al (2017) Guidelines for morpholino use in zebrafish. PLoS Genet 13(10): e1007000
6. Ablain J, Zon LI (2016) Chapter 9 - Tissue-specific gene targeting using CRISPR/Cas9. In: William Detrich H, Westerfield M, Zon LI (eds) Methods in cell biology. Academic, pp 189–202
7. Hwang WY et al (2013) Efficient genome editing in zebrafish using a CRISPR-Cas system. Nat Biotech 31(3):227–229
8. Yin L et al (2015) Multiplex conditional mutagenesis using transgenic expression of Cas9 and sgRNAs. Genetics 200(2):431–441
9. Kok FO et al (2015) Reverse genetic screening reveals poor correlation between morpholino-induced and mutant phenotypes in zebrafish. Dev Cell 32(1):97–108
10. Strykowski JL, Schech JM (2015) Effectiveness of recommended euthanasia methods in larval zebrafish (Danio rerio). J Am Assoc Lab Anim Sci 54(1):76–79
11. Wilson JM, Bunte RM, Carty AJ (2009) Evaluation of rapid cooling and tricaine methanesulfonate (MS222) as methods of euthanasia in zebrafish (Danio rerio). J Am Assoc Lab Anim Sci 48(6):785–789
12. Ferreira JM, Félix L, Jorge S, Monteiro SM, Olsson IAS, Valentim AM (2022) Anesthesia overdose versus rapid cooling for euthanasia of adult zebrafish. Zebrafish 19(4):148–159
13. Rosen JN, Sweeney MF, Mably JD (2009) Microinjection of zebrafish embryos to analyze gene function. JoVE 25:e1115
14. Thomas HR et al (2014) High-throughput genome editing and phenotyping facilitated by high resolution melting curve analysis. PLoS One 9(12):e114632
15. Hill JT et al (2014) Poly peak parser: method and software for identification of unknown indels using sanger sequencing of polymerase chain reaction products. Dev Dyn 243(12): 1632–1636
16. Gudnason H et al (2007) Comparison of multiple DNA dyes for real-time PCR: effects of dye concentration and sequence composition on DNA amplification and melting temperature. Nucleic Acids Res 35(19):e127–e127
17. Wang J et al (2022) Variable phenotypes and penetrance between and within different zebrafish ciliary transition zone mutants. Dis Model Mech 15(12)
18. Berger J, Sztal T, Currie PD (2012) Quantification of birefringence readily measures the level of muscle damage in zebrafish. Biochem Biophys Res Commun 423(4):785–788
19. Montandon M, Currie PD, Ruparelia AA (2021) Examining muscle regeneration in zebrafish models of muscle disease. J Vis Exp 167: e62071
20. Gupta VA et al (2013) Loss of catalytically inactive lipid phosphatase myotubularin-related protein 12 Impairs myotubularin stability and promotes centronuclear myopathy in zebrafish. PLOS Genetics 9(6):e1003583
21. Maves L (2014) Recent advances using zebrafish animal models for muscle disease drug discovery. Expert Opin Drug Discov 9(9):1033–1045
22. Singh J, Patten SA (2022) Modeling neuromuscular diseases in zebrafish. Front Mol Neurosci 15:1054573
23. Green MR, Sambrook J (2020) Isolation of total RNA from zebrafish embryos and adults. Cold Spring Harb Protoc 2020(1):pdb.prot101667
24. Casadei R et al (2011) Identification of housekeeping genes suitable for gene expression analysis in the zebrafish. Gene Expression Patterns 11(3):271–276
25. McCurley AT, Callard GV (2008) Characterization of housekeeping genes in zebrafish: male-female differences and effects of tissue type, developmental stage and chemical treatment. BMC Mol Biol 9(1):102
26. Rassier GT et al (2020) Evaluation of qPCR reference genes in GH-overexpressing transgenic zebrafish (Danio rerio). Sci Rep 10 (1):12692

Chapter 8

Preparation of Size-Defined PEG-Grafted Copolymers as a Polymeric Nanoruler for Size Optimization in Passive Targeting

Yukine Ishibashi, Mitsuru Naito, Chun Yin Jerry Lau, Yoshitsugu Aoki, and Kanjiro Miyata

Abstract

Size-dependent (or passive) targeting strategies are a major approach for designing nanomedicines (or nanocarriers) that can systemically deliver drugs to lesion sites as well as solid tumors. However, the optimal size range for passive targeting to various noncancerous lesion sites is unclear. Recently, we fabricated poly(ethylene glycol) (PEG)-grafted copolymers (gPEGs) as a size-tunable "polymeric nanoruler" to clarify the optimal size for systemic inflamed/noninflamed muscle tissue targeting. The hydrodynamic diameter (D_H) of gPEGs was controlled in the range from 11 nm to 32 nm by varying the molecular weights of grafted PEGs. The biodistribution analyses of gPEGs revealed that 11–15 nm sized gPEG achieved the highest accumulation in the skeletal muscle tissue of healthy mice, which was presumably determined by competition between the renal excretion efficiency and muscle tissue extravasation efficiency. Interestingly, muscular dystrophy model mice (*mdx* mice) showed two to three times higher levels of gPEG accumulation in the skeletal muscle tissue, compared with healthy mice, due to the enhanced extravasation of gPEGs in inflamed muscle tissues. The results demonstrated the utility of size-tunable gPEGs, i.e., polymeric nanoruler, for optimizing the passive targeting strategy. In the present study, we describe a detailed protocol for the preparation of gPEGs to disseminate the methodology of the polymeric nanoruler.

Key words Muscle targeting, Nanomedicine, Size-tuning, Duchenne muscular dystrophy, EPR-like effect

1 Introduction

Size-dependent (or passive) targeting strategies are a leading method for designing nanomedicines (or nanocarriers) that can systemically deliver drugs to target tissues in the body. This comes from the fact that the pore sizes of blood capillaries are highly dependent on organs and tissues. Indeed, nanocarriers with sizes of several tens to one hundred of nanometers have extensively been developed for passive tumor targeting [1, 2]. When administered

Yoshitsugu Aoki and Norio Motohashi (eds.), *Splice Switching: Methods and Protocols*, Methods in Molecular Biology, vol. 2963, https://doi.org/10.1007/978-1-0716-4738-7_8,

intravenously, such nanocarriers can avoid renal filtration having a size threshold of ~8 nm (or molecular weight of ~50,000) [3–5] and may extravasate via the leaky tumor vasculature for preferential accumulation in solid tumor tissues, which is well known as the enhanced permeability and retention (EPR) effect [1, 2, 6, 7]. Recently, it has been recognized that the EPR effect is considerably affected by tumor heterogeneity [8]; the tumor vasculature leakiness varies according to various tumor conditions even in the same cancer patient [8], and stromal tissues in fibrous tumors act as a barrier for nanocarriers to reach the cancer cell nest after extravasation [9]. To overcome such tumor heterogeneity, one plausible approach is downsizing nanocarriers. Smaller nanocarriers with a size of less than 30 nm more likely penetrate smaller tumor vascular pores as well as fibrotic stromal tissues [10–12]. These recent advances in nanomedicine research suggest that nanocarrier sizes need to be further optimized in a smaller range, e.g., 10–30 nm, for targeting appropriate cancer models that closely mimic patients with cancer. Furthermore, the optimal size range also remains unclear for passive targeting of various noncancerous lesion sites, including inflamed tissues.

To elucidate the optimal size range for cancer and inflamed tissue targeting, we recently constructed a size-tunable macromolecular platform as a "polymeric nanoruler" [13, 15]. A series of PEG-grafted copolymers (gPEGs) were synthesized by grafting poly(ethylene glycol)s (PEGs) with varying molecular weights (MWs) to the side chains of poly(aspartic acid) (PAsp) backbone, followed by fluorescent dye labeling for biodistribution studies. The sizes of the obtained gPEGs were successfully controlled with a hydrodynamic diameter (D_H) ranging from 11 to 32 nm and a narrow size distribution. When intravenously administered to Duchenne muscular dystrophy model (*mdx*) mice, gPEGs with a size of 11–32 nm achieved approximately 30-fold higher levels of accumulation in muscle tissues, compared to a simple PEG molecule with a size of 8 nm. Interestingly, the muscle tissue accumulation levels of 11–32 nm-sized gPEGs in *mdx* mice were two- to threefold higher than those in healthy mice, demonstrating enhanced extravasation, i.e., EPR-like effect, in the inflamed muscle tissues of *mdx* mice.

Herein, we describe a detailed methodology regarding the gPEG-based nanoruler study according to our previous work [13]: fabrication of size-defined gPEGs, determination of their D_H using fluorescence correlation spectroscopy, biodistribution (or pharmacokinetics) analysis, and intramuscular distribution of gPEGs.

2 Materials

2.1 Synthesis of gPEGs with Varying Sizes

Vacuum all flasks used for the synthesis of poly(β-benzyl-L-aspartate) (PBLA) over 1 h and repeat filling them with argon (Ar) gas and vacuuming three times before use. Flow Ar gas when the flask is opened and reagents are added.

1. *n*-Butylamine diluted solution: 1 mL dehydrated *n*-butylamine and 99 mL dehydrated dichloromethane (DCM) (100-fold dilution of dehydrated *n*-butylamine with dehydrated DCM). Store in a round-bottom flask under Ar atmosphere (*see* **Note 1**).
2. β-Benzyl-L-aspartate *N*-carboxy-anhydride (BLA-NCA): Store at −20 °C under Ar atmosphere (*see* **Note 2**).
3. Argon gas: 99.999% purity.
4. Plastic glove bag.
5. Gas-collecting balloon.
6. Membrane filter: poly(1,1,2,2-tetrafluoroethylene) (PTFE), 10.0 μm pore size.
7. Regenerated cellulose membranes (dialysis membranes): Wash inside the membrane prior to dialysis.
8. α-Azide-tetraethylene glycol-ω-amine (N_3-PEG_4-NH_2) diluted solution: 30 mM N_3-PEG_4-NH_2 in *N,N*-dimethylformamide (DMF). Weigh 100 mg N_3-PEG_4-NH_2 and add 12.7 mL of DMF. Store at −20 °C.
9. Cyanine5 functionalized with dibenzocyclooctyne (Cy5-DBCO) solution: 50 mg/mL Sulfo-Cyanine5 in dimethyl sulfoxide (DMSO). Store at −20 °C.
10. PD-10 purification buffer: 1 M NaCl in deionized water.

2.2 Size Exclusion Chromatography (SEC)

1. Eluent for PBLA: 50 mM lithium bromide (LiBr) in *N*-methyl-pyrrolidone (NMP). Add 4.34 g of LiBr to 1 L of NMP and stir overnight.
2. Elusion buffer: 10 mM phosphate buffer solution (pH 7.4), 0.15 M NaCl.

2.3 Fluorescence Correlation Spectroscopy (FCS) Measurements

1. FCS buffer: Dulbecco's phosphate-buffered saline (PBS) containing 0.05v% Tween 20 (PBS-T).
2. ATTO633 solution: 1 nM ATTO633 in PBS-T. Store at −20 ° C.
3. Cy5 solution: 10 nM Cyanine5 carboxylic acid in PBS-T. Store at −20 °C.

4. 384-well black plate with a clear bottom: Add 100 μL of Sigmacote solution to each well and remove the solution immediately. Dry the plate in the hood overnight.

2.4 Biodistribution Analysis

1. Homogenizing buffer: PBS-T.

3 Methods

3.1 Synthesis of gPEGs with Varying Sizes

1. Inside an Ar-filled plastic glove bag, weigh 500 mg of BLA-NCA and put it in a round-bottom flask (*see* **Note 3**). Dissolve BLA-NCA in 1 mL of dehydrated DMF and then add 9 mL of dehydrated DCM to adjust to a concentration of 50 mg/mL under an Ar atmosphere (*see* **Notes 4** and **5**).
2. Add the BLA-NCA solution to 0.18 mL of *n*-butylamine diluted solution with stirring under Ar atmosphere (*see* **Note 6**). After attaching the gas collection balloon to the reaction flask, stir the mixed solution at 35 °C for over 72 h under Ar atmosphere (*see* **Note 7**).
3. Collect a small amount of the reaction solution under Ar atmosphere and analyze the infrared (IR) spectrum for checking the disappearance of IR peaks at 910, 1780, and 1860 cm^{-1} (Fig. 1), which indicates the completion of BLA-NCA polymerization (*see* **Note 8**).
4. Dropwise, add the reaction solution into 150 mL of diethyl ether with vigorous stirring for 30 min. Collect the white precipitate (PBLA) by vacuum filtration using a membrane filter and dry in a vacuum (*see* **Note 9**).
5. Dissolve PBLA in the eluent for PBLA at a concentration of 1 mg/mL and conduct SEC analysis with the eluent for PBLA

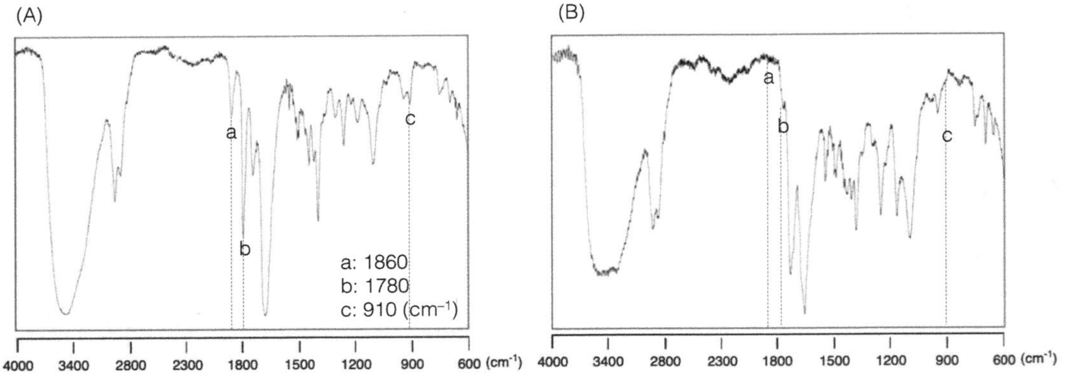

Fig. 1 IR spectra of (**a**) starting state and (**b**) completion state of β-benzyl-L-aspartate *N*-carboxy-anhydride (BLA-NCA) polymerization

to determine the molecular weight distribution (M_w/M_n) of PBLA using a calibration curve prepared by PEG standards (*see* **Note 10**).

6. Dissolve PBLA in DMSO-*d6* at a concentration of 10 mg/mL and determine the degree of polymerization of PBLA based on the peak intensity ratio of propylene protons in *n*-butylamine ($CH_3CH_2CH_2$–, δ = 0.7–1.4) to benzyl protons in PBLA ($C_6H_5CH_2$–, δ = 7.3) by ^{1}H NMR measurement (80 °C, 128 scans) (*see* **Note 10**).
7. Suspend 100 mg of PBLA in 2 mL acetonitrile and add 4.9 mL of 0.5 M NaOH (5 equiv. to BLA) to obtain PAsp by deprotecting the benzyl groups in PBLA. Stir the solution at room temperature until the solution turns clear.
8. Pour the reaction solution into the dialysis membrane (MWCO: 6–8 kDa) and dialyze against deionized water for half a day. Lyophilize the reaction mixture after dialysis to obtain PAsp with sodium salt (PAsp/Na) as a white powder. Dissolve PAsp/Na in D_2O at a concentration of 10 mg/mL and confirm the disappearance of benzyl peaks in the ^{1}H-NMR spectrum (25 °C, 128 scans).
9. Dissolve 50 mg of PAsp/Na in 12.5 mL of 50 mM $NaHCO_3$ aq. Add 1.216 mL of N_3-PEG$_4$-NH$_2$ diluted solution (0.1 equiv. of N_3-PEG$_4$-NH$_2$ to the carboxy group in PAsp), 69.96 mg of 1-(3-dimethylaminopropyl)-3-ethylcarbodiimide hydrochloride (EDC•HCl) (1 equiv. to the carboxyl group in PAsp), and 42.00 mg of *N*-hydroxysuccinimide (NHS) (1 equiv. to the carboxyl group in PAsp) to PAsp/Na solution and stir at room temperature for 3 h.
10. Pour the reaction solution into the dialysis membrane (MWCO: 6–8 kDa) and dialyze against 10 mM NaOH aq., 10 mM HCl aq., and deionized water. Each dialysis proceeds for half a day at 4 °C. Lyophilize the dialysis solution to obtain azide-modified PAsp as a free acid state.
11. Dissolve azide-modified PAsp in D_2O at a concentration of 10 mg/mL and determine the modification amount of the azide group based on the peak intensity ratio of the N_3-PEG$_4$-NH$_2$ protons ($N_3(CH_2CH_2O)_4CH_2CH_2NH$–, δ = 3.3–3.8) to methylene protons in PAsp ($CHCH_2CO$–, *δ* = 2.7–3.0) by ^{1}H-NMR (Fig. 2) (25 °C, 128 scans).
12. Dissolve 4 mg of azide-modified PAsp in 1 mL of DMSO. Prepare the same solution in four separate vials. Dissolve methoxy-PEG-NH$_2$ (PEG-NH$_2$) (MW: 2000, 5000, 12,000, and 20,000) in DMF at concentrations of 100 mg/mL separately (*see* **Note 11**).

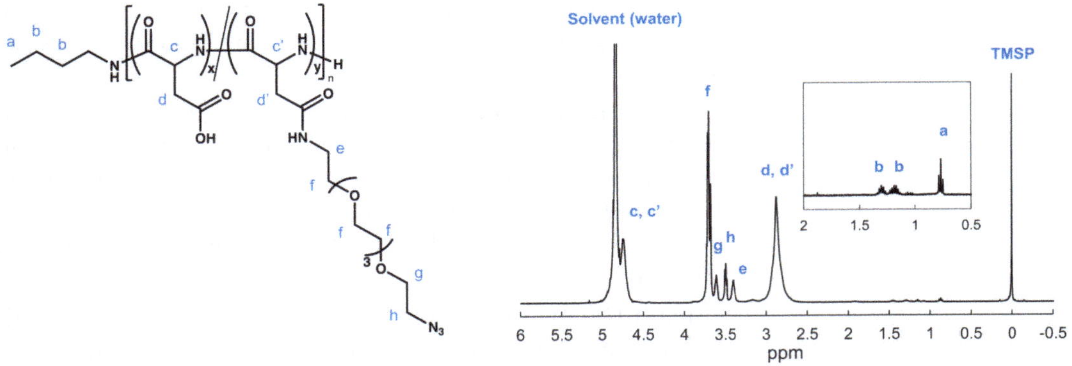

Fig. 2 Chemical structure and ^{1}H NMR spectrum of azide-modified PAsp. Only the α-form of PAsp is shown for the illustration of the chemical structure

13. Add 4.83 mg of EDC•HCl (0.95 equiv. to the carboxyl group in PAsp) and 3.36 mg of NHS (1.1 equiv. to the carboxyl group in PAsp) to the azide-modified PAsp solution and stir for a few minutes (*see* **Note 12**).
14. Add 1.063 mL of PEG-NH_2 with MW 2000, 2.656 mL of PEG-NH_2 with MW 5000, 6.375 mL of PEG-NH_2 with MW 12,000, and 10.625 mL of PEG-NH_2 with MW 20,000 to the PAsp solution (2 equiv. to the carboxyl group in PAsp) and stir it overnight to conjugate PEG-NH_2 to PAsp.
15. Mix the PEG-NH_2 solution and DMSO with the same volume ratio as above to prepare the reference PEG-NH_2 solution.
16. Conduct an SEC analysis of each reaction and reference solution with the elution buffer at a flow rate of 0.5 mL/min. Use a Superdex 75 Increase column for PEG with MW 2000 and a Superdex 200 Increase column for PEG with the other MWs (*see* **Note 13**).
17. Calculate the PEG introduction numbers to PAsp based on the PEG intensities detected by a refractive index (RI) detector. Compare the PEG intensity ratio of each PEG-NH_2 reference solution to the remaining PEG-NH_2 in the reaction solution (Fig. 3) (*see* **Note 14**).
18. Pour the reaction solution into a dialysis membrane (MWCO: 12–14 kDa) and dialyze against 10 mM NaOH aq. for half a day and deionized water for half a day at 4 °C. Lyophilize the reaction mixture after dialysis to obtain gPEG.
19. Remove the remaining unreacted PEG-NH_2 using preparative SEC equipped with a Superdex 200 prep-grade column with elution buffer (*see* **Note 15**).

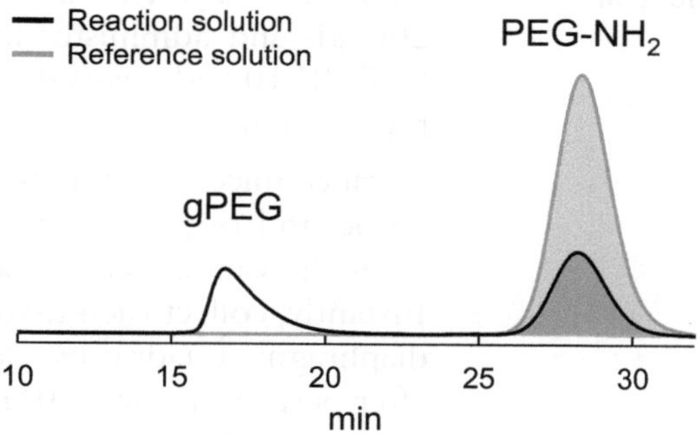

Fig. 3 SEC charts of gPEG solutions (black line) and the reference solutions (gray line) of PEG with molecular weight 2000. Peak areas of PEG-NH_2 are colored gray

20. Pour the reaction mixture into the dialysis membrane (MWCO: 12–14 kDa) and dialyze against deionized water for 1 day. Lyophilize the reaction mixture after dialysis.

3.2 Fluorescent Dye Labeling of gPEGs

1. Dissolve the purified gPEG in deionized water (25 mg/mL). Add the Cy5-DBCO solution to the gPEG solution (1 equiv. to gPEG) and subject the reaction solution to freeze–thaw treatment three times at −20 °C and 4 °C.
2. Pour the reaction solution into the dialysis membrane (MWCO: 12–14 kDa) and dialyze against deionized water, ethanol, deionized water, and 500 mM NaCl aq. for half a day for each and against deionized water for 1 day. Lyophilize the reaction solution.
3. Dissolve the obtained white powder in the PD-10 purification buffer and proceed with PD-10 purification to remove the remaining Cy5-DBCO using the PD-10 purification buffer (*see* **Note 16**).
4. Pour the eluted solution into the dialysis membrane (MWCO: 12–14 kDa) and dialyze against deionized water for 1 day. Lyophilize the purified solution to obtain Cy5-labeled gPEGs (Cy5-gPEGs) (*see* **Note 17**).

3.3 Fluorescence Correlation Spectroscopy (FCS) Measurements

1. Dissolve Cy5-*g*PEGs in PBS-T at a concentration of 25 nM and measure their diffusion time with a measurement time of 20 s and five repetitions at room temperature. Calibrate the FCS machine with the ATTO633 solution and measure the diffusion time of the Cy5 solution as a reference.
2. Calculate the D_H value of gPEGs from each diffusion time using the Stokes–Einstein equation. Use the diffusion coefficient value of Cy5 dye as a reference (360 μm^2/s at 25 °C) [14].

3.4 Biodistribution Analysis

1. Dissolve Cy5-*g*PEGs in PBS at a concentration of 10 μM in 200 μL and administer intravenously to C57/BL6 mice or C57/BL10-*mdx* (*mdx* mice) (7- to 9-week-old, male) through the tail vein.
2. Sacrifice mice under anesthesia using isoflurane to collect the blood and organs at the desired time points. Collect blood from the caudal vena cava under anesthesia using isoflurane. Instantly, collect each tissue (spleen, liver, lung, kidney, heart, diaphragm, quadriceps, gastrocnemius, and tibialis anterior) after perfusion using 10 mL of PBS via the left atrium. Rinse the collected tissues gently with PBS.
3. Weigh the collected organs and cut the liver into pieces of around 100 mg. Put each organ (a half piece for lung and kidney and ~ 100 mg piece for liver) into a homogenizing tube. Add 500 mL of PBS-T (1000 mL for quadriceps and gastrocnemius) together with a homogenizing metal cone.
4. Homogenize by shaking the tubes at 2500 rpm for 30 s, with 30 s intervals, and repeat shaking three times.
5. Transfer 100 μL of the homogenates and ten-fold-diluted blood with PBS to a 96-well black plate with a clear bottom.
6. Prepare a standard solution of injected samples in a half-log dilution series covering a range of 1% dose/well to 0.001% dose/well. Add 4 μL of administered sample to 196 μL of PBS-T as a 1% dose/well standard solution and dilute using PBS-T. Transfer standard solutions to a 96-well black plate with a clear bottom.
7. Measure the fluorescence intensities using a plate reader (excitation/emission = 630/675 nm).
8. Convert the obtained fluorescence intensities to the accumulation amounts per gram of tissue (dose%/g) (or dose%/mL for blood) (*see* **Note 18, 19**).
9. Determine blood circulation half-life by noncompartmental analysis and calculate the area under the curve (AUC) of each tissue based on the accumulated samples at each time point using the trapezoidal rule (*see* **Note 20**) (Fig. 4).

3.5 Fluorescence Immunohistochemistry

1. Place the collected quadriceps on a corkboard using tragacanth gum premixed with water. Freeze using precooled isopentane with liquid nitrogen.
2. Slice the frozen quadriceps into a 10-μm thick section using a cryomicrotome and dry overnight on a slide glass.
3. Perform fluorescence imaging for Cy5-gPEG and bright-field imaging using a fluorescence microscope (*see* **Note 21**).

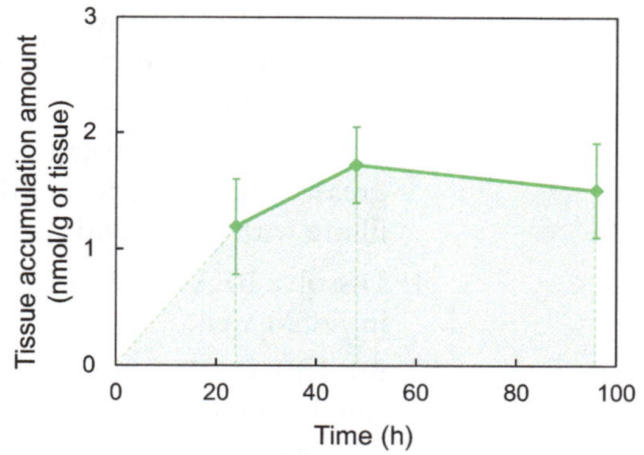

Fig. 4 Illustration of the AUC calculation method with the trapezoidal rule

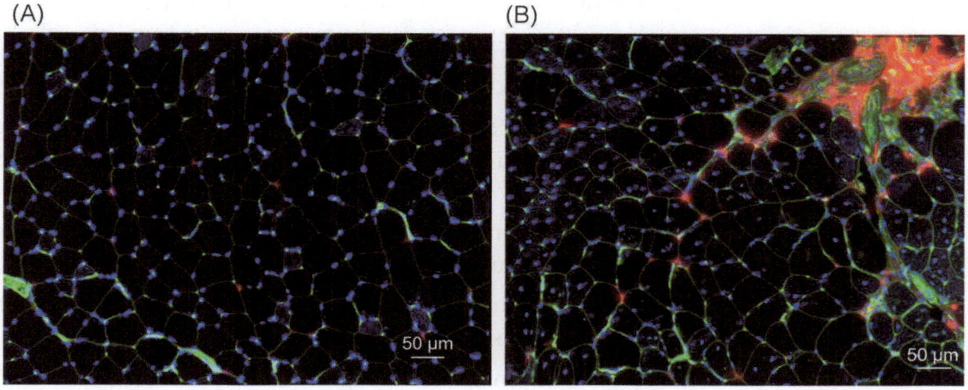

Fig. 5 Representative immunostaining section of the quadriceps of (**a**) C57/BL6 mice and (**b**) *mdx* mice 48 h after intravenous administration of Cy5-gPEG (red: Cy5-gPEG, blue: cell nucleus (DAPI), and green: plasma membrane (WGA-488))

4. Fix the section with 4% paraformaldehyde (PFA) in PBS at room temperature for 10 min and gently wash with PBS. Stain the cell membrane with wheat germ agglutinin (WGA) using Alexa Fluor 488 dye (WGA-488) (5 μg/mL) overnight at 4 °C in the dark and wash gently with PBS. Stain the nuclei with a mounting reagent containing DAPI.
5. Image the cell membrane (WGA-488) and nuclei (DAPI) at the same focal distance as the Cy5 channel and overlay these images (Fig. 5).

4 Notes

1. Distill DCM and *n*-butylamine with calcium hydride. Dehydrated-grade reagents are acceptable.

2. BLA-NCA is sensitive to moisture. Put BLA-NCA in a desiccator for an hour after taking it from a freezer until its temperature becomes ambient.
3. Put an electronic balance, a spatula, a weighing dish, a flask, grease, and a bottle of BLA-NCA in the glove bag. Repeat filling with Ar gas and vacuuming three times before weighing.
4. Dissolve BLA-NCA in DMF first. BLA-NCA does not dissolve in DCM well.
5. 50–100 mg/mL of the final BLA-NCA concentration is a suitable range.
6. Calculate the required amount of *n*-butylamine (or BLA-NCA) based on the following equation:

   (molar amount of *n*-butylamine) × (target degree of polymerization) × 1.1 = (molar amount of BLA-NCA).
7. CO_2 gas will generate from BLA-NCA during polymerization.
8. Extend the reaction time when the IR peaks appear.
9. Prepare 15–20 excess amounts of diethyl ether to the reaction solution. Put a reprecipitation beaker in ice to improve the yield.
10. If PBLA does not dissolve in the eluent for PBLA, heat and sonicate them until the sample turns clear.
11. If PAsp does not dissolve in DMSO, heat and sonicate them until the PAsp solution turns clear.
12. Do not add an excess amount of NHS, which leads to side reactions.
13. Depending on the detector sensitivity, dilute the reference and reaction solutions with DMSO or DMF, followed by SEC analysis.
14. Calculate the introduction number of PEG to PAsp as follows: (Introduction number of PEG) = {(Reference PEG area)—(Remaining PEG are)}/(Reference PEG area) × 2 × (Polymerization degree of PAsp).
15. Ultrafiltration cassettes are also useful for removing the remaining PEGs.
16. PD-10 column tends to be clogged when applying a high-MW gPEG solution. When the PD-10 column does not work well, dilute the solution to reduce the viscosity.
17. Store at 4 °C at 20 μM or a higher concentration in PBS.
18. Calculate the accumulation amount per gram of tissues from the measured fluorescence intensities (FI) as follows.
    (a) Calculate the weight of tissues in 100 μL of homogenate (g in well).

(g in well) = (weight of tissue in homogenizing tube, g) × 100/{(weight of tissue in homogenizing tube, g) + (added amount of homogenizing buffer, mL)}

(b) Subtract autofluorescence intensity.

(FI/g in well, gPEGs without autofluorescence) = (FI/g in well, gPEGs)—(FI/g in well, non-treated)

(c) Convert to dose%/g of tissue using the fluorescence intensity of the standard solution.

(dose%/g of tissue) = (FI/g in well, gPEGs without autofluorescence)/(FI/%).

19. Convert dose%/g of tissues to the accumulated molar amount of gPEGs to each tissue (accumulated molar amount/g of tissue) as follows:

(accumulated molar amount/g of tissue) = (dose%/g of tissue) × (administered molar amount, 2 nmol).

20. Calculate the AUC value as illustrated in Fig. 4.

21. As Cy5-gPEGs are readily leached from the section during the wash procedure, imaging for Cy5-gPEGs should be performed before fixation.

Acknowledgments

This study was financially supported by the Japan Science and Technology Agency (JST) under the Center of Innovation (COI)-NEXT Program, Japan Society for the Promotion of Science (JSPS) under Grants-in-Aid for Scientific Research (KAKENHI Grant Number: 18 K18379 for MN), Japan Agency for Medical Research and Development (AMED) under Grant Number JP22ak0101134 for KM, and the Intramural Research Grant (Grant Number 2–6) for National Center of Neurology and Psychiatry (NCNP).

Conflict of Interest The authors have no conflicts of interest to declare that are relevant to the content of this chapter.

References

1. Cabral H, Miyata K, Osada K et al (2018) Block copolymer micelles in nanomedicine applications. Chem Rev 118:6844–6892. https://doi.org/10.1021/acs.chemrev.8b00199
2. Mi P, Miyata K, Kataoka K et al (2021) Clinical translation of self-assembled cancer nanomedicines. Adv Ther 4:2000159. https://doi.org/10.1002/adtp.202000159
3. Seymour LW, Duncan R, Strohalm J et al (1987) Effect of molecular weight of N-(2-hydroxypropyl)methacrylamide copolymers on body distribution and rate of excretion after subcutaneous, intraperitoneal, and

intravenous administration to rats. J Biomed Mater Res 21:1341–1358. https://doi.org/10.1002/jbm.820211106

4. Yamaoka T, Tabata Y, Ikada Y (1994) Distribution and tissue uptake of poly(ethylene glycol) with different molecular weights after intravenous administration to mice. J Pharm Sci 83: 601–606. https://doi.org/10.1002/jps.2600830432
5. Choi HS, Liu W, Misra P et al (2007) Renal clearance of quantum dots. Bat Biotechnol 25: 1165–1170. https://doi.org/10.1038/nbt1340
6. Matsumura Y, Maeda H (1986) A new concept for macromolecular therapeutics in cancer chemotherapy: mechanism of tumoritropic accumulation of proteins and the antitumor agent Smancs. Cancer Res 46:6387–6392
7. Maeda H, Wu J, Sawa T, Matsumura Y, Hori K (2000) Tumor vascular permeability and the EPR effect in macromolecular therapeutics: a review. J Control Release 65:271–284. https://doi.org/10.1016/S0168-3659(99)00248-5
8. Hansen AE, Petersen AL, Henriksen JR et al (2015) Positron emission tomography based elucidation of the enhanced permeability and retention effect in dogs with cancer using copper-64 liposomes. ACS Nano 9:6985–6995. https://doi.org/10.1021/acsnano.5b01324
9. Kano MR, Bae Y, Iwata C et al (2007) Improvement of cancer-targeting therapy, using nanocarriers for intractable solid tumors by inhibition TGF-β signaling. Proc Natl Acad Sci USA 104:3460–3465. https://doi.org/10.1073/pnas.0611660104
10. Cabral H, Matsumoto Y, Mizuno K et al (2011) Accumulation of sub-100 nm polymeric micelles in poorly permeable tumours depends on size. Nat Nanotechnol 6:815–823. https://doi.org/10.1038/nnano.2011.166
11. Watanabe S, Hayashi K, Toh K et al (2019) In vivo rendezvous of small nucleic acid drugs with charge-matched block catiomers to target cancer. Nat Commun 10:1894. https://doi.org/10.1038/s41467-019-09856-w
12. Kim HJ, Yi Y, Kim A, Miyata K (2018) Small delivery vehicles of siRNA for enhanced cancer targeting. Biomacromolecules 19:2377–2390. https://doi.org/10.1021/acs.biomac.8b00546.[]
13. Naito M, Watanuki Y, Toh K et al (2022) Size-tunable PEG-grafted copolymers as a polymeric nanorular for passive targeting muscle tissues. J Control Release 347:607–614. https://doi.org/10.1016/j.jconrel.2022.05.030
14. Loman A, Dertinger T, Koberling F et al (2008) Comparison of optical saturation effects in conventional and dual-focus fluorescence correlation spectroscopy. Chem Phys Lett 459:18–21. https://doi.org/10.1016/j.cplett.2008.05.018
15. Ishibashi Y, Naito M, Watanuki Y et al (2024) Size-dependent glioblastoma targeting by polymeric nanoruler with prolonged blood circulation. Bioconjug Chem 35(8):1154–1159. https://doi.org/10.1021/acs.bioconjchem.4c00235

Chapter 9

Splice Modulation Studies Using Reporter Mice

Katsura Minegishi, Tsukasa Tominari, and Yoshitsugu Aoki

Abstract

Exon-skipping therapy for Duchenne muscular dystrophy (DMD) recovers muscle function by inducing shorter but functional dystrophin protein using antisense oligonucleotides (ASOs) that modulate the pre-mRNA splicing of *DMD* transcripts to restore the disrupted *DMD* reading frame. Next-generation ASO should have appropriate sequences and modifications that induce exon skipping and ensure efficient delivery to target tissues. Conventional methods involve analyzing biopsied tissues using RT-PCR or immunoblotting to evaluate exon skipping or the expression of induced proteins. However, these methods are laborious and time-consuming, resulting in a low throughput. In this chapter, we present a novel evaluation method using reporter mice. Exon skipping induces the expression of fluorescent proteins enabling noninvasive monitoring. This in vivo method allows for a high-throughput evaluation of the efficacy of ASO-induced exon skipping.

Key words Duchenne muscular dystrophy, Exon skipping, Reporter mouse, In vivo imaging system (IVIS), Fluorescence, Luciferase

1 Introduction

Duchenne muscular dystrophy (DMD) is an X-linked genetic disorder caused by the loss or reduction of dystrophin protein, resulting in progressive muscle wasting in the skeletal, cardiac, and respiratory muscles [1]. To date, there has been no effective treatment for DMD, but several therapeutic approaches have recently been investigated [2–4]. Among these, exon-skipping therapy has great potential to effectively treat DMD. This approach uses antisense oligonucleotides (ASOs) to modulate pre-mRNA splicing of *DMD* transcripts to restore the disrupted *DMD* reading frame and induce a shorter but functional dystrophin protein (Fig. 1a) [5–7]. To effectively induce exon skipping in the target gene and deliver it to the target tissues correctly, appropriate ASO sequences and modifications should be selected [8]. Conventionally, exon skipping and the delivery of ASOs are evaluated by RT-PCR using tissues biopsied from mice injected with ASOs. Furthermore, the

Yoshitsugu Aoki and Norio Motohashi (eds.), *Splice Switching: Methods and Protocols*, Methods in Molecular Biology, vol. 2963, https://doi.org/10.1007/978-1-0716-4738-7_9,

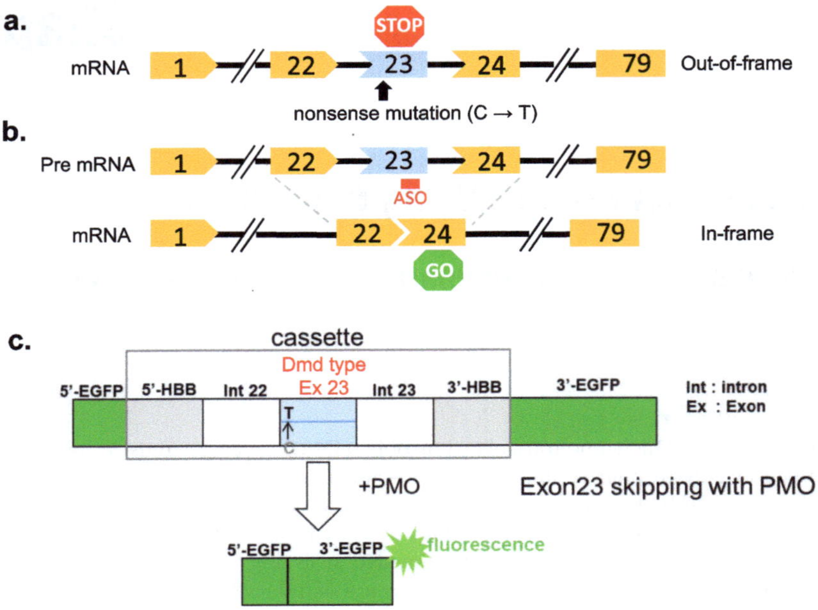

Fig. 1 In vivo screening analysis using the mdx-exon23/enhanced green fluorescence protein (EGFP) vector. (**a**) Nonsense mutation of exon 23 (C > T) caused frameshift (out-of-frame), leading to the production of a truncated dystrophin protein. (**b**) Antisense oligonucleotides' (ASOs) targeted exon 23 skipping converts the out-of-frame transcript into an in-frame transcript, allowing the synthesis of short but functional dystrophin. (**c**) Structure of the EGFP-exon 23 cassette, used to report dystrophin exon skipping, is depicted. The EGFP gene is divided into two regions, 5′-EGFP and 3′-EGFP, with the insertion of a mouse dystrophin genomic region containing the exon 23 and neighboring introns (Int 22 and Int 23). Skipping exon 23 with ASOs restores the EGFP open-reading frame, resulting in the expression of EGFP proteins

expression of the induced short but functional dystrophin is examined by immunoblotting. However, it is difficult to screen ASOs on a large scale using these methods because they are laborious and time-consuming. In addition, it is difficult to evaluate exon skipping and delivery over time using these methods.

Here, we introduce a new method using a reporter mouse that enables noninvasive monitoring of exon skipping. Using this method, exon skipping can be evaluated with a high throughput, which enables effective large-scale screening [9].

2 Materials

2.1 Design of the Reporter Vector and Creation of the Mouse Line

The detailed design of the reporter vector can be found in Subheading 3.1 of this chapter.

The detailed method for creating transgenic mice is described in the book "Manipulating the Mouse Embryo: A Laboratory Manual" [10].

2.2 Intramuscular Injection of ASO for the Induction of Exon Skipping

- 8–10-week-old reporter mice.
- Isoflurane.
- Hair removal cream and a shaver.
- Sterile insulin syringes 29G × 1/2 needles.
- Morpholino antisense oligo (mEx23D) 10 mM (GeneTools).

2.3 Intravenous Injection of Antisense Morpholino in Mice (Retro-Orbital Injection)

- 8–10-week-old reporter mice.
- Isoflurane.
- Sterile insulin syringes 29G × 1/2 needles.
- Morpholino antisense oligo (mEx23D) 10 mM (GeneTools).
  (5′-GGCCAAACCTCGGCTTACCTGAAAT-3′) [9].
- Mouse feed for fluorescence imaging (chlorophyll-free) (D10001, RESEARCH DIETS).

2.4 Evaluation of Exon Skipping by Using the In Vivo Imaging System (IVIS)

- IVIS imaging system (PerkinElmer, Waltham, MA, USA).
- Isoflurane.
- Oxygen cylinder.

3 Methods

3.1 Construction of Reporter Vectors and Creation of Mouse Lines

The noninvasive analysis of exon skipping in the *DMD* gene requires the use of fluorescent protein or luciferase reporter mice. Several research groups have created their own reporter mice to evaluate exon skipping [9, 11]. Here, we present the detailed description of our mdx-EGFP reporter mice, which were specifically developed by our team. Our ASO drug-screening system incorporates an enhanced green fluorescent protein (EGFP)-based reporter vector driven by a CMV early enhancer/chicken β actin (CAG) promoter [12]. Within the reporter vector, a fragment of exon 23 of *DMD* carrying a nonsense mutation, which is the same as that in the mdx mice, is inserted into a split EGFP (EGFP-exon23/pCAGGS). When ASOs induce skipping of exon 23, the EGFP reading frame is restored, leading to the expression of the EGFP protein (Fig. 1c) [9].

1. For the construction of vectors, purified PCR products are assembled using the NEB HiFi DNA assembly kit or the in-fusion HD cloning kit following the manufacturer's protocol. The vectors can be edited either by restriction enzyme digestion and ligation or recombination methods.
2. To generate transgenic (Tg) mice, each transgene (the linearized EGFP-exon 23/pCAGGS reporter vector) is microinjected into the pronucleus of fertilized eggs obtained by

crossing C57BL/6 J female and male mice. The eggs are then transferred to pseudopregnant mothers. Tg mice are genotyped using genomic PCR analysis, and subsequently, stable Tg lines are established [10].

3.2 Intramuscular Injection of PMO Solutions for the Induction of Exon Skip

1. Load insulin syringes using needles with 10 μg of the oligonucleotides diluted in saline (50 μL) (*see* **Note 1**).
2. Anesthetize the mice using isoflurane.
3. Remove the hair around the tibialis anterior (TA) muscle using a hair removal cream or a shaver.
4. Inject the PMO solution into the TA muscle using a 29G insulin syringe. Insert the needle tip from the distal side of the TA muscle and advance it toward the center, reaching the proximal side of the TA muscle. Inject the solution while gradually withdrawing the needle to allow the solution to disperse evenly within the muscle.
5. Return the mice to the cage.

3.3 Intravenous Injection of Antisense Morpholino in Mice (Retro-Orbital Injection)

1. Prepare PMO solution and withdraw 100 μL into a 0.5 mL insulin syringe.
2. Anesthetize the mice using isoflurane.
3. Place the needle in the retrobulbar space.
4. Return the mice to the cage.

When a luciferase reporter is used, a chlorophyll-free feed is recommended to reduce tissue autofluorescence (*see* **Note 2**, Fig. 4b) [13].

3.4 Evaluation of Exon Skipping by Using the In Vivo Imaging System (IVIS)

Several days (e.g., 1–2 weeks) following PMO injection, EGFP expression is induced as a result of exon skipping. The location and intensity of fluorescence indicate the location and extent of exon skipping. Fluorescence analysis is performed using IVIS (PerkinElmer, Waltham, MA, USA), which enables monitoring and quantification of diffused light emitted from a source using a cooled CCD camera. This noninvasive method allows for repeated evaluation of the same mouse to assess changes over time.

In cases where tissue autofluorescence interferes with monitoring, the surrounding area of the monitoring region is shaded with black paper, or mice are fed chlorophyll-free diet to minimize autofluorescence effects.

1. Set up the anesthesia machine following the instruction manual.
2. Launch the control software "Living Image 2.6" for the IVIS.

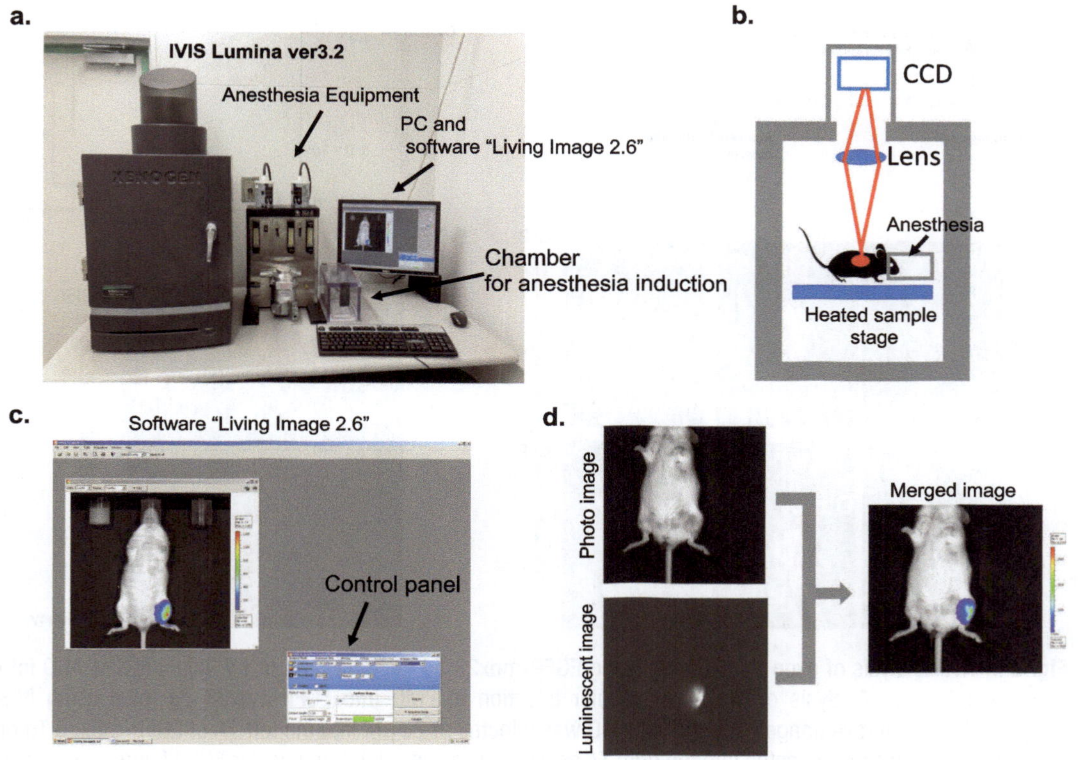

Fig. 2 The in vivo imaging system (IVIS). (**a**) An overall view of the imaging IVIS. (**b**) Schematic representation of the IVIS. (**c**) User interface of the software used for system operation. (**d**) Fluorescent (or luminescent) images and a photographic image are acquired and automatically merged to create a composite image

Click on the "Initialize" button in the IVIS acquisition control panel and wait for the CCD camera to reach a cooled temperature of −90 °C.

3. Note that the IVIS acquisition control panel is accessible only on a PC workstation connected to the IVIS (*see* **Note 3**, Fig. 2c). The available options in the control panel will vary based on the selected imaging mode (luminescent or fluorescent) and acquisition mode. To detect EGFP, select a filter set for EGFP (excitation and emission), and set exposure time to 5–10 s with an iris setting of F = 2. To detect luciferase, set exposure time to 60 s with an iris setting of F = 1.
4. In the case of using luciferase reporter mice, inject intraperitoneally with 100 μL of d-luciferin solution or AkaLumine-HCl (TokeOni) [14–16].
5. Place the mouse into the anesthesia chamber.
6. After the mouse is anesthetized, immediately transfer the mouse to the IVIS stage. Position the mouse on its back with the nose in the nose cone. To detect EGFP, shade the areas

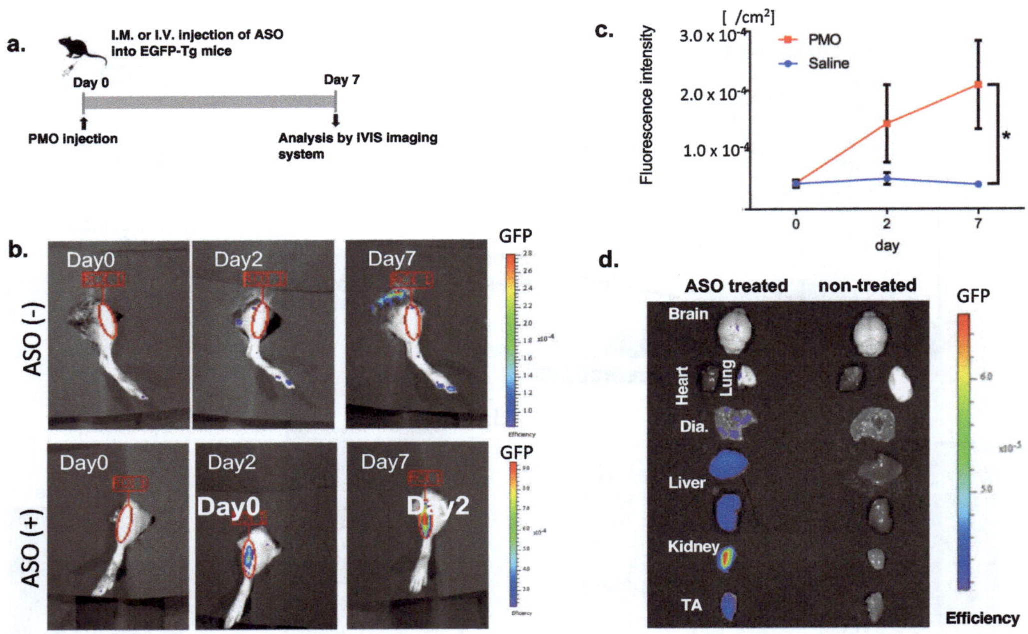

Fig. 3 In vivo analysis of exon 23 skipping using EGFP-mdx23 Tg mice. (**a**) I.M. or I.V. injections of ASO into EGFP-Tg mice. (**b**) Analysis of the intramuscular injection of PMO into EGFP-mdx23 Tg mice using IVIS imaging. Representative images are shown. PMO was injected at 20 μM into the left TA of EGFP-mdx23 Tg or WT mice, and saline was injected into the right TA as a negative control. Intramuscular EGFP fluorescence was analyzed on days 0, 2, and 7, using an in vivo Imaging System imaging (IVIS). (**c**) EGFP fluorescence intensity increased over time following ASO injection. (**d**) Representative images of ex vivo analysis using IVIS images after systemic injection of PMO into EGFP-mdx23 Tg mice. The mice were injected with 640 mg/kg body weight of ASO. The left column shows the muscles and organs of ASO-treated mice, and the right column shows those of saline-treated mice

outside the monitoring region with a black paper to minimize the effect of autofluorescence, if necessary (Fig. 4a).

7. Push the "Acquire" button in the control panel to capture the image. The system will acquire both the fluorescent (or luminescent) image and the photographic image, saving them automatically. The merged image of these images will be generated automatically (Fig. 2d).
8. Use the ROI tool in the edit panel to select the region of interest (ROI) in the merged image. The intensity value of the ROI will be calculated automatically (Fig. 3a, b).

   <Supplement>

   To evaluate the fluorescence (or luminescence) of internal organs, extract the organs from the mouse and place them on the IVIS stage. The fluorescence (or luminescence) signal can be evaluated using the aforementioned process described above (Fig. 3c).

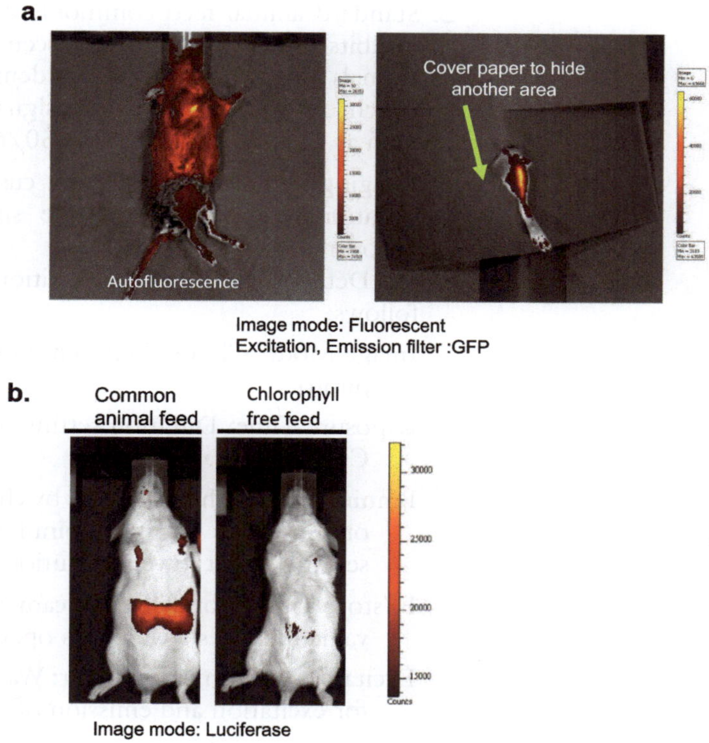

Fig. 4 Autofluorescence in tissue affects imaging. (**a**) In cases where tissue autofluorescence interferes with EGFP fluorescence, the area outside the monitored region is shaded with black paper. (**b**) Detection of luciferase signal by the IVIS may be affected by autofluorescence in the gut and stomach, which originates from the animal feed. The use of chlorophyll-free feed can help suppress the autofluorescence caused by chlorophyll, thereby improving imaging quality

9. Once the evaluation is complete, transfer the mouse back to the cage.
10. If necessary, collect tissues from the mouse for further analysis. Extract RNA or proteins from the tissues to perform RT-PCR or immunoblotting, allowing for comparison with the results obtained from the IVIS evaluation.

4 Notes

1. The PMO injection volume for each TA muscle in adult mice (>8 weeks old) ranges from 50 to 100 μL. The specific volume depends on the age and body weight of the analyzed mice. For the gastrocnemius (GAS) or quadriceps femoris (QF) muscle of adult mice, the estimated injection volume is 150 μL for GAS and 100 μL for QF.

2. Standard animal feed commonly contains chlorophyll, which exhibits maximum autofluorescence at 680 nm. Tissues of animals fed with this feed also demonstrate autofluorescence, which can interfere with the signal from fluorescent probes such as Cy 5.5, Alexa Fluoro 650/680, and EGFP.
3. Imaging conditions should be customized according to the situation. In cases where the signal is low, increase the exposure time.

   Details of the imaging conditions in the control panel are as follows:

   Image Mode: Choose between luminescence, fluorescence, or photo.

   Exposure Time: Duration setting for photon counting by the CCD camera.

   Binning: Adjust the sensitivity by changing the virtual pixel size of the CCD. A higher binning value results in a higher sensitivity but lower resolution.

   F/stop: Corresponds to the camera's iris opening. A smaller value indicates a wider iris opening.

   Excitation and Emission Filter: Wavelength setting of the filter for excitation and emission of light.

Acknowledgments

We thank all the members of the Department of Molecular Therapy, NCNP, for their technical assistance, especially Yoko Okubo and Yuko Hara.

Conflict of Interest The authors have no conflicts of interest to declare that are relevant to the content of this chapter.

References

1. Hoffman EP, Brown RH, Kunkel LM (1987) Dystrophin: the protein product of the duchenne muscular dystrophy locus. Cell 51(6): 919–928
2. Tominari T, Aoki Y (2022) Clinical development of novel therapies for Duchenne muscular dystrophy—current and future. Neurol Clin Neurosci 11(3):111–118
3. Mah JK, Korngut L, Dykeman J, Day L, Pringsheim T, Jette N (2014) A systematic review and meta-analysis on the epidemiology of Duchenne and Becker muscular dystrophy. Neuromuscul Disord 24(6):482–491
4. Crone M, Mah JK (2018) Current and emerging therapies for Duchenne muscular dystrophy. Curr Treat Options Neurol 20(8): 31
5. Siva K, Covello G, Denti MA (2014) Exon-skipping antisense oligonucleotides to correct missplicing in neurogenetic diseases. Nucleic Acid Ther 24(1):69–86
6. Aartsma-Rus A, Straub V, Hemmings R et al (2017) Development of exon skipping therapies for Duchenne muscular dystrophy: a critical review and a perspective on the outstanding issues. Nucleic Acid Ther 27(5):251–259

7. Echevarría L, Aupy P, Goyenvalle A (2018) Exon-skipping advances for Duchenne muscular dystrophy. Hum Mol Genet 27(R2):R163–R172
8. Khvorova A, Watts JK (2017) The chemical evolution of oligonucleotide therapies of clinical utility. Nat Biotechnol 35(3):238–248
9. Hara Y, Mizobe Y, Inoue YU et al (2020) Novel EGFP reporter cell and mouse models for sensitive imaging and quantification of exon skipping. Sci Rep 10(1):10110
10. Nagy A, Gertsenstein M, Vintersten K, Behringer R (2002) Manipulating the mouse embryo: a laboratory manual. Cold Spring Harbor Laboratory Press, New York
11. Amoasii L, Li H, Zhang Y et al (2019) In vivo non-invasive monitoring of dystrophin correction in a new Duchenne muscular dystrophy reporter mouse. Nat Commun 10(1):4537
12. Hitoshi N, Ken-ichi Y, Jun-ichi M (1991) Efficient selection for high-expression transfectants with a novel eukaryotic vector. Gene 108(2):193–199
13. Weagle G, Paterson PE, Kennedy J, Pottier R (1988) The nature of the chromophore responsible for naturally occurring fluorescence in mouse skin. J Photochem Photobiol B 2(3):313–320
14. Iwano S, Sugiyama M, Hama H et al (2018) Single-cell bioluminescence imaging of deep tissue in freely moving animals. Science 359(6378):935–939
15. Kuchimaru T, Iwano S, Kiyama M et al (2016) A luciferin analogue generating near-infrared bioluminescence achieves highly sensitive deep-tissue imaging. Nat Commun 7(1):11856
16. Iwano S, Obata R, Miura C et al (2013) Development of simple firefly luciferin analogs emitting blue, green, red, and near-infrared biological window light. Tetrahedron 69(19):3847–3856

Chapter 10

Clinical Variant Databases and Machine Learning Prediction Supporting Genomic Medicine

Mayumi Kamada and Yosuke Kawai

Abstract

Genomic medicine, which provides diagnosis and treatment decisions based on genomic variations, has been implemented in clinical practice and become more accessible. The clinical interpretation of genomic variations detected through genome analysis is crucial in genomic medicine. Clinical interpretation often utilizes genomic databases and effects predictions by machine learning. However, it has been noted that the genomic data accumulated thus far exhibit biases toward certain ethnic groups, leading to efforts to enhance genomic data diversity across various regions. While diverse and highly accurate machine learning methods have been developed for coding region variants, prediction methods for noncoding region variants still need to be improved in accuracy and applicable region. This chapter outlines the basic flow of genomic medicine and how detected variants are interpreted, introducing disease variant databases to advance genomic medicine and machine learning methods for predicting the effect of noncoding region variants.

Key words Genomic medicine, Clinical interpretation of variants, Database, Machine learning

1 Introduction

"Genomic medicine," which involves diagnosis, treatment, and prevention based on the variations in our genome, is beginning to be implemented in clinical practice. In genomic medicine, clinical interpretations are made for variants detected by mainly next-generation sequencing. These interpretations guide the decisions for diagnosis and treatment strategies. Clinical interpretation incorporates each case's clinical background, including family history, the frequency of the variant in the population, and its association with diseases, as shared in academic papers, case reports, and public databases. Therefore, accumulating and sharing knowledge about variants have been dated as a global challenge.

Most variants accumulated so far have been detected through genomic tests such as target sequencing and whole exome sequencing (WES), focusing mainly on variants in the gene coding regions.

Yoshitsugu Aoki and Norio Motohashi (eds.), *Splice Switching: Methods and Protocols*, Methods in Molecular Biology, vol. 2963, https://doi.org/10.1007/978-1-0716-4738-7_10,

However, with the recent reduction in whole genome sequencing (WGS) costs, large-scale WGS analysis projects are being conducted worldwide. WGS analysis has revealed various structural variations, including diverse variants in noncoding regions. Variants that affect the splicing donor and acceptor within 2 bp of exon–intron boundaries, known as canonical splice site (CSS) variants, are widely recognized as "loss of function (LoF)" variants, potential candidates for disease diagnosis [1]. Meanwhile, it has been found that "near-splice" regions within tens of base pairs of exon–intron boundaries also contain well-conserved splicing motifs, suggesting the presence of potentially damaging noncoding variants [2–4]. Furthermore, variants at branch points or deep intronic variants more than 100 bp away from exon–intron boundaries that inhibit splicing have also been suggested to contribute to disease onset [5].

As global projects continue to advance whole-genome analysis, the detection of many noncoding variants is inevitable. Clarifying their molecular functions and relevance to diseases is crucial. However, it has been pointed out that many noncoding variants other than CSS, such as near-splice variants, are not reported in public clinical databases [3]. For variant interpretation, sharing knowledge and developing computational prediction methods that comprehensively utilize this knowledge will be key.

This chapter begins with a brief introduction to the flow of genomic medicine with clinical sequencing. It then follows an overview of databases that serve as repositories for accumulating and sharing knowledge and the current state of prediction methods using machine learning.

2 Genome Analysis for Genomic Medicine

How do we identify the presence or absence of variants and interpret them clinically? Figure 1 outlines the general flow of analysis in genomic medicine (Fig. 1). The process begins with extracting genomic DNA from the patient's blood and tissue samples and then determining the nucleotide sequences using next-generation sequencers. The fragments of sequences (called "read") decoded by the sequencer are then compared with a reference sequence of humans to detect genomic differences. This comparison process with the reference sequence is called mapping. Based on the mapping results, a list of variants is created by applying variant call algorithms while utilizing existing information. Although there is a wide variety of algorithms for detecting variants, and details are omitted here, the Genome Analysis Toolkit (GATK) [6] is often used for germline variants. Mutect2 [7] is commonly employed for the detection of somatic variants. The detected variants are annotated with gene information using annotation tools such as VEP [8] or ANNOVAR [9], which help narrow down important variants'

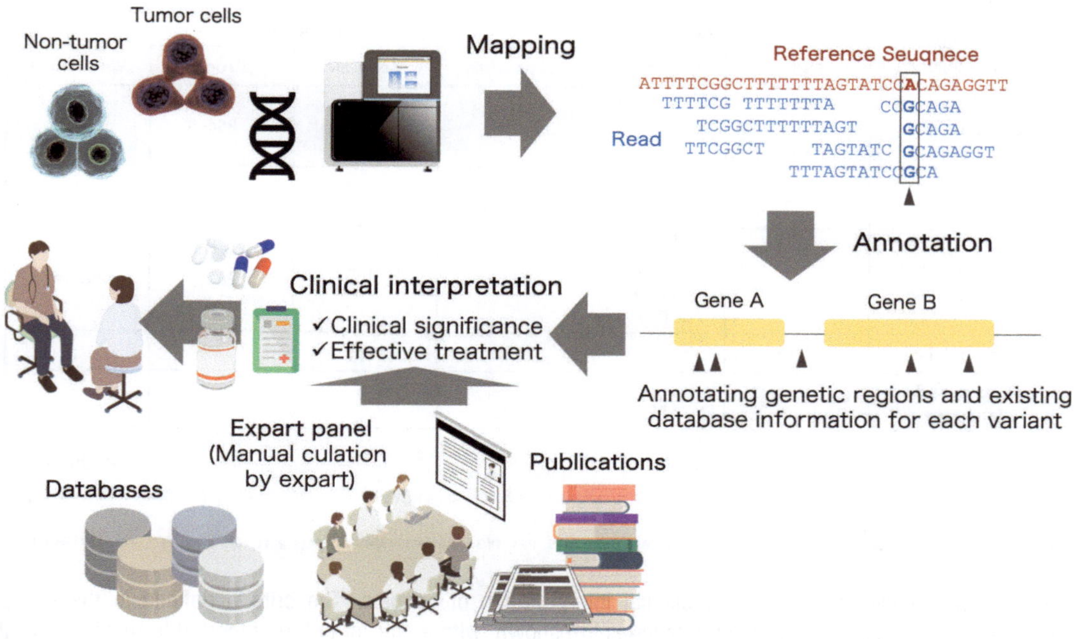

Fig. 1 Flow of clinical sequencing for genomic medicine. Detection of genomic variants using next-generation sequencers and processes to return results to patients

findings. Then, clinical interpretations of the variants are made based on the patient's detailed information and existing public database and literature, which are then used to decide on the diagnosis and treatment strategies. When analyzed with sequencers, blood or other nontumor tissues are examined for germline testing, while cancer cells are the focus for somatic gene testing, which targets tumor tissues. However, in somatic gene testing, nontumor and tumor tissues are often analyzed together to differentiate between inherited variants and those acquired later in life.

3 Clinical Interpretation of Variants

The most critical step in genomic medicine is the clinical interpretation of detected variants. This interpretation is carried out by comprehensively examining the clinical background of the case, including family history, the variant's frequency within the population, and its association with diseases, as shared in academic papers, case reports, and public databases. Since clinical interpretation requires standards, professional societies are advancing the creation of guidelines.

For Mendelian genetic diseases, the clinical genetic testing guidelines created by the American College of Medical Genetics and Genomics (ACMG) and the Association for Molecular

Criteria/ Data category	Benign		Pathogenic			
	Strong	Supporting	Supporting	Moderate	Strong	Very Strong
Population data	BA1 BS1 BS2			PM2	PS4	
Segregation data & Allelic data	BS4	BP2 BP5		PM3★ PM6	PS2	
Computational & Predictive data		BP1 BP3 BP4★ BP7	PP3★	PM4 PM5★	PS1	PVS1★
Functional data	BS3★		PP2	PM1	PS3★	
Other data & databases		BP6	PP4 PP5			

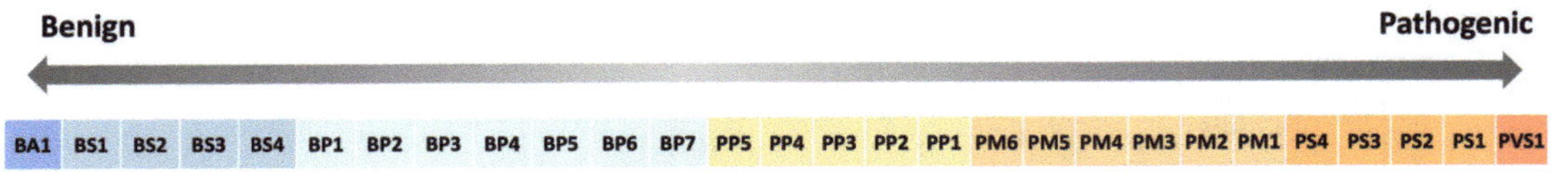

Fig. 2 ACMG guidelines criteria for clinical interpretation of variants. The criteria defined in the ACMG guidelines and their respective data categories are shown, with a star for criteria reported to require special consideration when applied to noncoding region variants. *ACMG* American College of Medical Genetics and Genomics, *PVS* Pathogenic Very Strong, *PS* Pathogenic Strong, *PM* Pathogenic Moderate, *PP* Pathogenic Supporting, *BA* Benign Stand-Alone, *BS* Benign Strong, *BP* Benign Supporting

Pathology (AMP), known as the ACMG guidelines, are used as the global standard [1]. The ACMG guidelines define categories based on the strength of evidence supporting the pathogenicity of a variant. Figure 2 shows defined categories (Fig. 2). For example, if a variant results in a loss of function (LOF) and is a null variant (such as a nonsense variant leading to a stop codon or a frameshift variant that changes the reading frame) in a gene believed to be involved in the onset mechanism of the disease, it is considered to meet the "Pathogenic Very Strong 1 (PVS1)" evidence of support. Conversely, if a variant is reported with an allele frequency of more than 5% in major population allele frequency databases, it meets the criteria for "evidence for benign (Benign) Stand-Alone (BA1)." The ACMG guidelines assess which category criteria each piece of evidence for a variant meets and then determine its final pathogenicity in five levels ("Pathogenic," "Likely pathogenic," "Benign," "Likely benign," and "Uncertain significance") based on the combination of category criteria.

Many of the criteria in ACMG guidelines assume coding region variants, especially missense variants, where a single nucleotide change results in substituting one amino acid for another. Therefore, it has been pointed out that special consideration and adaptation are necessary for noncoding region variants [10], especially for criteria marked with a star in Fig. 2. Particularly, in the interpretation of noncoding region variants, the limited availability of in silico scores (computationally predicted scores) or the necessary additional information (such as epigenomic data to define candidate regulatory elements) is reported as one of the significant barriers to applying the guidelines.

4 Databases for Clinical Interpretations

For clinical interpretations, conducting a comprehensive investigation of variants and gathering evidence are necessary. Therefore, databases related to disease association and treatment information play a crucial role.

ClinVar [11], operated by the National Center for Biotechnology Information, is widely used as a disease-related variant database. Launched in 2013 under the National Institutes of Health, ClinVar has registered about 2.8 million unique variation records from over 2700 contributors, including research institutions and genetic testing companies, as of March 2024. ClinVar contains data primarily in four domains: variant data, associated conditions/phenotypes, clinical interpretations, and their evidence. ClinVar provides the reliability of clinical interpretations and their evidence with a review status indicated by stars, making it a trusted database in the field of genomic medicine.

However, most genomic and disease research to date has been conducted on populations of European ancestors. Thus, concerns have emerged over the gathered genomic data's failure to fully reflect the diversity of populations worldwide, which could obstruct the accurate understanding of diseases and worsen health inequalities among various ethnic groups [12]. The ACMG guidelines use population allele frequencies in their determination of pathogenicity. The Genome Aggregation Database (gnomAD) [13], which provides frequency information for populations based on large-scale genomic data, reports that about 60% of its aggregated data comes from people with European ancestry. In comparison, sequences from people with African ancestry comprise less than 10% [14]. This disparity in data is problematic because it has become clear that variants affecting disease can differ among ethnic groups. For instance, the frequency of somatic mutations in driver genes causing non-small-cell lung cancer has been reported to vary between East Asian and Western populations [15]. The proportion of mutations in the CFTR gene, commonly associated with cystic fibrosis, differs between populations of European and African ancestors, leading to diagnostic confusion [16]. It has been pointed out that such biases in genomic data can indeed affect the outcomes of genomic medicine. A study comparing the results of panel tests for hereditary cancers among different ethnic groups has reported that Asian and Hispanic populations have a higher proportion of variants with uncertain clinical significance compared to Europeans [17].

Therefore, efforts are being made to increase the diversity of genomic data across different populations. In Africa, projects like The African Genome Variation Project and Human Heredity and Health in Africa (H3Africa) actively promote the accumulation of genomic data and research [18, 19]. Similarly, in Japan, the "Program for an Integrated Database of Clinical and Genomic Information" initiated by the Japan Agency for Medical Research and Development (AMED) in 2016 serves as a pioneer for consolidating genomic information in East Asia. The project has developed "MGeND" (Medical Genomics Japan Database [20], targeting diseases such as cancer, rare and intractable diseases, infectious diseases, dementia, and hearing loss to aggregate clinical genomic information.

MGeND (https://mgend.ncgm.go.jp/) was launched in March 2018 as an "open-access database" and has been updated. It covers monogenic and polygenic diseases, allowing cross-disease search of variant data. Because the primary genomic analysis method varies by the disease area, the variant data handled in MGeND are diverse (Table 1). Genome-wide association studies (GWAS) are widely used to research polygenic diseases. Studies focusing on diseases involving the immune system, such as infectious diseases, often concentrate on the polymorphism (allele types) of HLA (human leukocyte antigen). This protein plays a central role in the immune response. Therefore, MGeND includes SNV, short indel, SV, and CNV data as treated in ClinVar, as well as variant sets obtained from GWAS analysis and HLA allele frequency data.

A comparison with variants registered in ClinVar revealed that about 65% were unique to MGeND (as of January 2024). In addition to variants with agreeing clinical interpretations, many variants with differing interpretations were also included (Fig. 3).

Table 1
Number of submitted variations in MGeND (MGeND assigns disease area labels according to the data provider's research area. The table shows the number of variants provided by each disease area group)

Data type/disease area	Variant (Sequenced data)	GWAS data	HLA allele data
Cancer	160,760	–	–
Rare/intractable disease	36,331	584	–
Infectious disease	1417	155,100	1803
Dementia	12,864	761	–
Others	11,638	10,779	–

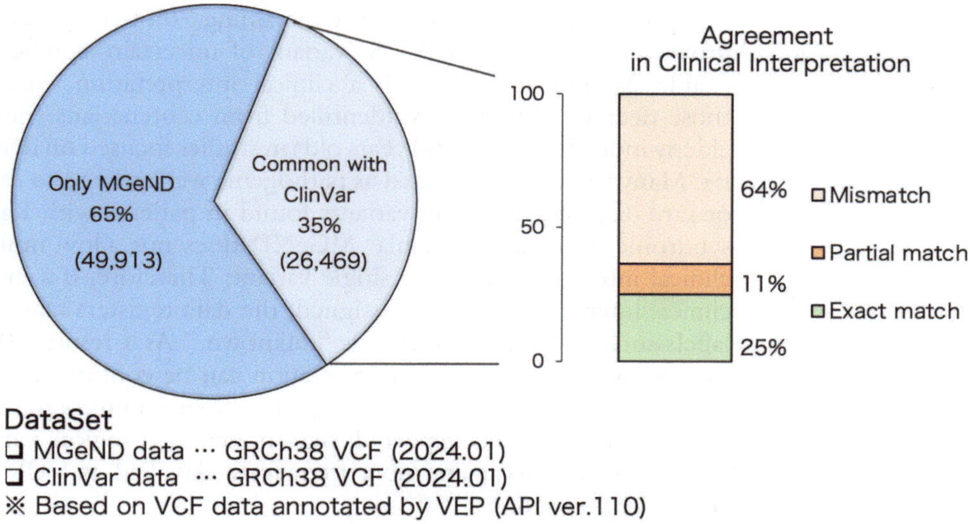

Fig. 3 Comparison with MGeND variants and ClinVar variants. About 70% of the unique variants were not registered in ClinVar but only in MGeND. Of the shared variants, many of them had disagreeing clinical interpretation labels

Table 2
Number of variants in MGeND regarding splicing (Each entry in MGeND (GRCh38 VCF format data) is annotated with VEP v.110. Entries annotated as splice donor/acceptor or splice region variants are listed)

Clinical significane in MGeND	Number of entries
Pathogenic	1742
Benign	1653
Likely pathogenic	209
Likely benign	215
Uncertain significance	685
Other	40
Not provided	8761

This highlights the importance of efforts to accumulate genomic data across different ethnic groups.

Focusing on variants related to splicing in MGeND, annotation was performed using VEP on VCF format data (hg38) comprising 142,800 entries (defined uniquely by submitter, genomic position, change, and disease name). Entries annotated with splice-related variants (splice acceptor/donor/splice region) totaled 13,305, with approximately 10% (7075) being unique variants related to splice regions, covering all disease areas included in MGeND (Table 2). About 30% of these splice-related variants had clinical interpretations attached, while the remaining 70% mainly were not provided or classified as VUS (variant of uncertain significance) (Table 2). For variants with a clinical interpretation, many of those deemed benign were identified from centenarians (healthy elderly individuals over 100 years old) in studies focused on dementia. Many variants interpreted as pathogenic were related to cancer or rare diseases, including variants found in patients with Kabuki syndrome or Fanconi anemia. MGeND does not allow multiple clinical interpretations for a single variant. Therefore, if a unique clinical interpretation is not assigned, the data registers may apply labels such as "not provided" or "unknown." As a result, 70% of variants without a clinical interpretation can be considered either having an undetermined clinical significance or uninterpretable.

It has also been reported that noncoding region variants registered in ClinVar are more frequently classified as VUS than coding region variants [10].

The primary reason for the lack of clinical significance is insufficient information necessary for clinical interpretation. It is essential to continue sharing and accumulating research findings. Furthermore, using computational prediction methods based on machine learning, which will be introduced in Subheading 5, is expected to facilitate interpretation.

5 Prediction of the Effect of Variants

Many predictive methods have been developed to assess the molecular impact of coding region variants. In particular, many prediction methods for missense variants have been developed with machine learning, and the ACMG guidelines recommend using these prediction results as supportive information. Large-scale sequence data and deep-learning-based methods have recently enabled more accurate impact predictions [21, 22]. As mentioned earlier, the predictive methods applicable to noncoding region variants are limited. However, several tools have been developed to predict the impact of variants for splicing [23]. It is important to note that this does not imply a recommendation for using the mentioned tools.

SpliceAI [24] is a tool that employs a deep neural network to predict splicing sites from primary sequences. Specifically, it uses a 10,000-nucleotide pre-mRNA sequence as input to obtain the probability that each nucleotide position is a splice acceptor, a splice donor, or neither. The accuracy of SpliceAI has been reported to be high compared to multiple tools [23]. SpliceAI was trained using protein-coding gene annotations from GENCODE [25] (information on donor–acceptor pairs and splice junctions) combined with splice junction information based on RNA-Seq datasets from the Gene and Tissue Expression (GTEx) atlas [26]. Inputting sequences with variants enables predicting splicing changes caused by variants based on highly accurate splice site predictions (95% accuracy on the test set).

MMSplice (modular modeling of splicing) [27] constructs separate neural network models for assessing exons, introns, and splice sites (donor/acceptor), combining their outputs to predict the impact of variants on splicing. Modularizing rather than a single large neural network allows for utilizing large-scale experimental datasets obtained for different aspects of splicing in each module. The splice site module was trained using GENCODE annotation data. The exon and intron modules were trained using data from millions of random short sequences through a massively parallel reporter assay (MPRA) [28]. Outputs from each module are combined using a linear model to calculate scores evaluating the impact on exon skipping, donors, and acceptors. This linear model was trained using high-throughput assay data (Vex-seq dataset) [29] that quantifies the impact of natural exon variants on their exon splicing. Their approach enables highly accurate predictions, and MMSplice won first place in the exon-skipping prediction category of the fifth challenge of the Critical Assessment of Genome Interpretation group (CAGI5, https://genomeinterpretation.org/). A disease association (pathogenicity) prediction model has also been constructed by combining the outputs of each module using a logistic regression model.

CADD-Splice [30] integrates splicing scores from SpliceAI and MMSplice into The Combined Annotation Dependent Depletion (CADD), a prediction tool highly regarded for its reliability in the domain of rare diseases [31]. CADD is widely used to predict the effects of genome-wide variants and integrates prediction scores from various tools by weighting them. It employs a logistic regression model trained on more than 15 million evolutionary-derived variants as negative (benign) data and a matching set of simulated variants as positive (deleterious) data. CADD-Splice demonstrated that incorporating splicing impact scores into CADD significantly enhances the prediction accuracy for splicing-related mutations.

Beyond these, numerous predictive models for splice impact have been developed, including methods designed explicitly for diseases. For Mendelian diseases, there are methods like Genomiser [32] and NCBoost [33]. Genomiser is an analysis framework for noncoding variants associated with Mendelian diseases, the key to which is the Regulatory Mendelian Mutation (ReMM) score integrated into the framework. ReMM is a machine learning predictive model based on Random Forest [34], trained using a manually curated set of Mendelian disease variants. Genomiser uses the ReMM score and information from several other data sources to comprehensively evaluate pathogenicity. NCBoost trains a machine learning model called XGBoost [35] on a dataset including 737 pathogenic noncoding SNVs across 283 genes, curated manually from sources such as the Human Gene Mutation Database

(HGMD-DM) [36], ClinVar [11][Landrum MJ, NAR, 2016], and Smedley 2016 [32], enabling more accurate predictions of noncoding variant impacts in Mendelian diseases.

regBase is a prediction method targeted at cancer research [37]. regBase is a database that aggregates prediction scores from 13 noncoding variant effect predictors, such as CADD and ReMM, for all possible substitutions of SNV (8,575,894,770 substitutions) in the human reference sequence. Additionally, it integrates scores from 10 tools (including DeepSEA [38] and deltaSVM [39]) predicted for 38,248,779 biallelic variants obtained from the 1000 Genome Project phase 3, using scores from all 23 tools for biallelic variants to build a database named regBase Common. The authors constructed three regBase models (regBase_REG, regBase_PAT, and regBase_CAN) tailored to three prediction targets using these scores as explanatory variables and XGBoost [35]. Specifically, regBase_CAN, aimed at cancer, uses COSMIC recurrent regulatory mutations as training data to predict noncoding variants that could be cancer drivers. In a validation using a dataset constructed with COSMIC V84 [40] targeting Somatic eGene [41] obtained from expression quantitative trait loci (eQTL) analysis, regBase_CAN significantly outperformed comparative methods with an accuracy (AUC = 0.94).

The production of valuable experimental data by medical and biological scientists, along with the efforts of researchers in bioinformatics and related fields, has led to the development of various predictive methods. However, it is difficult to say that these methods are promising in predicting effects for all noncoding regions. In particular, attention must be paid to the limited data for training machine learning models and the potential for data bias. The paper on NCBoost pointed out the issue of region bias (limited data does not allow for equally accurate predictions across all noncoding regions) [33].

6 Conclusion

In this chapter, we focused on the flow of clinical interpretation in genomic medicine, highlighting the importance of databases and computational prediction methods using machine learning. As noncoding region variants' importance and complexity become more detailed, their clinical interpretation becomes increasingly crucial. Although many researchers' efforts have advanced numerous studies and the development of prediction methods, the bottleneck in clinical interpretation remains a significant challenge. However, as genomic analysis projects progress, data production, accumulation, and sharing are expected to improve the accuracy of predictions currently challenged by machine learning and eliminate biases.

Conflict of Interest

The authors have no conflicts of interest to declare that are relevant to the content of this chapter.

References

1. Richards S, Aziz N, Bale S et al (2015) Standards and guidelines for the interpretation of sequence variants: a joint consensus recommendation of the American College of Medical Genetics and Genomics and the Association for Molecular Pathology. Genet Med 17:405–424
2. Rivas MA, Pirinen M, Conrad DF et al (2015) Human genomics. Effect of predicted protein-truncating genetic variants on the human transcriptome. Science 348:666–669
3. Lord J, Gallone G, Short PJ et al (2019) Pathogenicity and selective constraint on variation near splice sites. Genome Res 29:159–170
4. Zhang S, Samocha KE, Rivas MA et al (2018) Base-specific mutational intolerance near splice sites clarifies the role of nonessential splice nucleotides. Genome Res 28:968–974
5. Nieminen TT, Pavicic W, Porkka N et al (2016) Pseudoexons provide a mechanism for allele-specific expression of APC in familial adenomatous polyposis. Oncotarget 7:70685–70698
6. Poplin R, Ruano-Rubio V, DePristo MA, et al (2018) Scaling accurate genetic variant discovery to tens of thousands of samples bioRxiv 201178
7. Benjamin D, Sato T, Cibulskis K, et al (2019) Calling somatic SNVs and Indels with Mutect2. bioRxiv 861054
8. McLaren W, Gil L, Hunt SE et al (2016) The Ensembl variant effect predictor. Genome Biol 17:1–14
9. Wang K, Li M, Hakonarson H (2010) ANNOVAR: functional annotation of genetic variants from high-throughput sequencing data. Nucleic Acids Res 38:1–7
10. Ellingford JM, Ahn JW, Bagnall RD et al (2022) Recommendations for clinical interpretation of variants found in non-coding regions of the genome. Genome Med 14:73
11. Landrum MJ, Lee JM, Benson M et al (2016) ClinVar: public archive of interpretations of clinically relevant variants. Nucleic Acids Res 44:D862–D868
12. Sirugo G, Williams SM, Tishkoff SA (2019) The missing diversity in human genetic studies. Cell 177:26–31
13. Karczewski KJ, Francioli LC, Tiao G et al (2020) The mutational constraint spectrum quantified from variation in 141,456 humans. Nature 581:434–443
14. Karczewski K (2017) The genome aggregation database (gnomAD). MacArthur Lab https://macarthurlab.org/2017/02/27/the-genome-aggregation-database-gnomad/. Accessed 15 Aug 2023
15. Kohno T, Nakaoku T, Tsuta K et al (2015) Beyond ALK-RET, ROS1 and other oncogene fusions in lung cancer. Transl Lung Cancer Res 4:156–164
16. Stewart C, Pepper MS (2017) Cystic fibrosis in the African diaspora. Ann Am Thorac Soc 14: 1–7
17. Caswell-Jin JL, Gupta T, Hall E et al (2018) Racial/ethnic differences in multiple-gene sequencing results for hereditary cancer risk. Genet Med 20:234–239
18. Gurdasani D, Carstensen T, Tekola-Ayele F et al (2014) The African genome variation project shapes medical genetics in Africa. Nature 517:327–332
19. H3Africa Consortium, Rotimi C, Abayomi A et al (2014) Research capacity. Enabling the genomic revolution in Africa. Science 344: 1346–1348
20. Kamada M, Nakatsui M, Kojima R et al (2019) MGeND: an integrated database for Japanese clinical and genomic information. Hum Genome Var 6:4–8
21. Frazer J, Notin P, Dias M et al (2021) Disease variant prediction with deep generative models of evolutionary data. Nature 599:1–5
22. Cheng J, Novati G, Pan J et al (2023) Accurate proteome-wide missense variant effect prediction with AlphaMissense. Science 381: eadg7492
23. Rowlands C, Thomas HB, Lord J et al (2021) Comparison of in silico strategies to prioritize rare genomic variants impacting RNA splicing for the diagnosis of genomic disorders. Sci Rep 11:1–11
24. Jaganathan K, Kyriazopoulou Panagiotopoulou S, McRae JF et al (2019) Predicting splicing from primary sequence with deep learning. Cell 176:535–548.e24
25. Harrow J, Frankish A, Gonzalez JM et al (2012) GENCODE: the reference human

genome annotation for the ENCODE project. Genome Res 22:1760–1774

26. Ardlie K, DeLuca DS, Segrè A et al (2015) The genotype-tissue expression (GTEx) pilot analysis: multitissue gene regulation in humans. Science 348:648–660
27. Cheng J, Nguyen TYD, Cygan KJ et al (2019) MMSplice: modular modeling improves the predictions of genetic variant effects on splicing. Genome Biol 20:48
28. Rosenberg AB, Patwardhan RP, Shendure J, Seelig G (2015) Learning the sequence determinants of alternative splicing from millions of random sequences. Cell 163:698–711
29. Adamson SI, Zhan L, Graveley BR (2018) Vex-seq: high-throughput identification of the impact of genetic variation on pre-mRNA splicing efficiency. Genome Biol 19:71
30. Rentzsch P, Schubach M, Shendure J, Kircher M (2021) CADD-splice-improving genome-wide variant effect prediction using deep learning-derived splice scores. Genome Med 13:31
31. Rentzsch P, Witten D, Cooper GM et al (2019) CADD: predicting the deleteriousness of variants throughout the human genome. Nucleic Acids Res 47:D886–D894
32. Smedley D, Schubach M, Jacobsen JOB et al (2016) A whole-genome analysis framework for effective identification of pathogenic regulatory variants in Mendelian disease. Am J Hum Genet 99:595–606
33. Caron B, Luo Y, Rausell A (2019) NCBoost classifies pathogenic non-coding variants in Mendelian diseases through supervised learning on purifying selection signals in humans. Genome Biol 20:32
34. Breiman L (2001) Random forests. Mach Learn 45:5–32
35. Chen T, Guestrin C (2016) XGBoost: a scalable tree boosting system. In: Proceedings of the 22nd ACM SIGKDD international conference on knowledge discovery and data mining. Association for Computing Machinery, New York, pp 785–794
36. Stenson PD, Mort M, Ball EV et al (2014) The human gene mutation database: building a comprehensive mutation repository for clinical and molecular genetics, diagnostic testing and personalized genomic medicine. Hum Genet 133:1–9
37. Zhang S, He Y, Liu H et al (2019) regBase: whole genome base-wise aggregation and functional prediction for human non-coding regulatory variants. Nucleic Acids Res 47:e134
38. Zhou J, Troyanskaya OG (2015) Predicting effects of noncoding variants with deep learning-based sequence model. Nat Methods 12:931–934
39. Lee D, Gorkin DU, Baker M et al (2015) A method to predict the impact of regulatory variants from DNA sequence. Nat Genet 47: 955–961
40. Tate JG, Bamford S, Jubb HC et al (2019) COSMIC: the catalogue of somatic mutations in cancer. Nucleic Acids Res 47:D941–D947
41. PCAWG Transcriptome Core Group, Calabrese C, Davidson NR et al (2020) Genomic basis for RNA alterations in cancer. Nature 578:129–136

Chapter 11

In Vivo Evaluation of CTG Repeat-Affected Muscle Pathology in a Myotonic Dystrophy Model Mouse Using Electromyography and Fluorescence In Situ Hybridization

Yusuke Echigoya, Katsura Minegishi, Masayuki Nakamori, and Yoshitsugu Aoki

Abstract

Myotonic dystrophy type 1 (DM1) is the most common form of muscular dystrophy in adults, caused by an expanded CTG repeat in the *DMPK* gene. Patients with DM1 develop multisystemic symptoms including myotonia, which manifests as delayed relaxation of muscles following voluntary contraction. This myotonic phenomenon can be assessed by electromyography (EMG), where it presents as a prolonged action potential duration, referred to as myotonic discharge. In both patients and mouse models, the expanded CUG repeat RNA transcribed from the mutated gene aberrantly accumulates in nuclei. This accumulation can be visualized using fluorescence in situ hybridization (FISH). The abnormal RNA accumulation leads to the dysregulation of alternative splicing in specific genes, resulting in disease symptoms that include myotonia and progressive muscle wasting. Given the pathogenesis, RNA-targeted therapeutics, such as antisense oligonucleotides, represent one of the promising approaches for DM1 treatment. This chapter describes a method for conducting needle EMG with a portable device, as well as a FISH method for a representative mouse model of DM1, the HSA^{LR} mouse harboring an expanded CTG repeat. These methods can be applied to evaluate the therapeutic efficiency of emerging therapies in DM1 model mice.

Key words Myotonic dystrophy type 1, CTG repeat, HSA^{LR} mice, Electromyography, Myotonic discharge, FISH, RNA foci

1 Introduction

Myotonic dystrophy type 1 (DM1) is the most common form of adult-onset muscular dystrophy, inherited in an autosomal dominant fashion with a prevalence of 1 in every 2100 births [1]. Individuals with DM1 develop a range of multisystemic manifestations, including myotonia, progressive muscle weakness, insulin resistance, easy fatigability, cardiac conduction defects, and cognitive

Authors Yusuke Echigoya and Katsura Minegishi have equally contributed to this chapter.

Yoshitsugu Aoki and Norio Motohashi (eds.), *Splice Switching: Methods and Protocols*, Methods in Molecular Biology, vol. 2963, https://doi.org/10.1007/978-1-0716-4738-7_11,

and behavioral dysfunctions [2]. Among these, the most common initial symptom is myotonia, characterized by the impairment of relaxation of skeletal muscles following voluntary contraction, significantly impacting patients' quality of life. Although several clinical trials are ongoing, there is currently no treatment for DM1 [3, 4].

DM1 is caused by an aberrant expansion of the CTG-trinucleotide repeat in the 3′ untranslated region (UTR) of the *DMPK* gene [5]. The mutated gene generates and ectopically accumulates CUG repeat RNA in nuclei, which can be visualized as toxic RNA foci using fluorescence in situ hybridization (FISH) with a repeat RNA target probe. The accumulation of toxic RNA foci disrupts RNA-binding protein functions, i.e., the CUG repeat RNA sequestrates muscle-blind-like (MBNL) proteins and upregulates CUGBP/Elav-like family (CELF) proteins. The disruption of MBNL and CELF functions due to the CUG repeat accumulation results in the misregulation of alternative splicing events in various genes, such as *chloride channel 1* (*CLCN1*) and *insulin receptor*, consequently developing multisystemic symptoms. Given the pathomechanism, RNA-targeted therapeutics, such as antisense oligonucleotides, for reducing the expression levels of CUG repeat RNA [3] represent a promising approach for DM1.

When assessing the in vivo effectiveness of candidate therapies, it is crucial to evaluate functional improvements such as the reduction of myotonia. Myotonia, characterized by delayed muscle relaxation, is identifiable through electromyography (EMG) by observing prolonged action potential duration, known as myotonic discharge, following needle insertion into the affected muscles of individuals with DM1 and certain murine models, such as the HSA^{LR} mouse model [6]. The HSA^{LR} mouse model has the *human skeletal actin* gene (*HSA* or *ACTA1*) with a long ~220 CTG repeat inserted in the 3′ untranslated region. Developed by Dr. Charles A. Thornton's group at the University of Rochester in 2000 [7], this model has been widely employed in DM1 research as a valuable tool for analyzing the muscle pathology affected by CUG repeat RNA toxicity [8]. The HSA^{LR} mice have been demonstrated to mimic the molecular pathomechanism of DM1 muscles, such as nuclear RNA foci and subsequent aberrant splicing in various genes [9].

Consequently, this HSA^{LR} mouse model allows for the analysis of myotonic discharge on needle EMG as indicative of myotonia [10]. In the mouse model, myotonia is primarily attributed to the misregulation of *Clcn1* splicing, which results from the nuclear accumulation of the expanded CUG repeat RNA, mirroring the situation observed in patients. Detecting the RNA foci in nuclei is critical for evaluating the efficiency of candidates for DM1 therapies in mouse models [11]. This chapter describes the methods of needle EMG using a portable device and the FISH method in

HSA^{LR} mice. The presented methods, essential in the preclinical development of DM1 therapeutics, can be applied to other mouse models of DM1.

2 Materials

2.1 HSA^{LR} Transgenic Mice and Wild-Type Mice

1. FVB/N-Tg(HSA*LR)20bCath/J mice (HSA^{LR} mice) are available from The Jackson Laboratory (https://www.jax.org/strain/032031) (*see* **Note 1**).
2. Wild-type FVB/N mice are available from The Jackson Laboratory (https://www.jax.org/strain/001800) or other suppliers.

2.2 Genotyping PCR (See Note 2)

1. Isoflurane.
2. Rodent anesthesia machine with an anesthesia induction chamber.
3. 70% ethanol, disinfectant.
4. Mouse-ear puncher (about 1.5 mm in diameter).
5. DNA extraction kit, e.g., DNeasy Blood & Tissue Kit (Qiagen).
6. PCR kit to amplify a DNA region with a high GC content, e.g., KOD FX Neo kit (Toyobo).
7. Primers designed to amplify the CTG repeat region in *ACTA1*. Synthesize them with reverse-phase cartridge purification or more. Sample primers are listed in Table 1.
8. 2% agarose gel in 1 × TAE buffer: For a prestained method, include a DNA stain, e.g., SYBR Safe DNA gel stain (Invitrogen) or ethidium bromide, at an adequate concentration, according to the manufacturer's instructions.
9. PCR product purification kit.
10. Equipment: an instrument for photometric DNA quantification, PCR thermal cycler, electrophoresis device, and UV transilluminator or similar imaging systems.

Table 1
Sample genotyping and sequencing primers to detect the CTG repeat region in the *ACTA1* gene of the HSA^{LR} transgene (GenBank accession no. M20543.1)

Name	Sequence 5′ to 3′	Amplicon size
hACTA1_g3403-24_Forward	GCCATCGTAAACTGACACAGTG	91 bp + (CTG)n
hACTA1_g3473-93_Reverse	GTTCCACAGGGCTTTGTTTCG	

2.3 Needle EMG Using a Portable Device

1. Isoflurane.
2. Rodent anesthesia machine with an anesthesia induction chamber.
3. Compact electric hair clipper for pets or depilatory cream.
4. Electromagnetic wave-shielding sheet.
5. Portable EMG device with a recording function for sound and waveform, e.g., Neuropack n1 (catalog no. MEM-8301, Nihon Kohden) with accessory devices to connect a concentric needle electrode (female and male connectors, catalog nos. BM-901B and BM-325C, respectively, Nihon Kohden).
6. 27-gauge concentric needle electrode, disposable (e.g., catalog no. NM-325C, Nihon Kohden).
7. Ground electrode (28-gauge monopolar needle electrode, disposable, e.g., catalog no. NM-225B, Nihon Kohden).
8. Disposable pocket warmer wrapped in a paper towel or a thermostatic heating plate.
9. Recovery cage, e.g., a small plastic container with a lid.

2.4 Collecting and Embedding Skeletal Muscle Tissues into Gum on a Cork Disc

1. Rodent anesthesia machine with an anesthesia induction chamber and isoflurane.
2. Micro-dissecting scissors, e.g., eye scissors.
3. Fine-tip dissecting forceps.
4. Tragacanth gum (*see* **Note 3**).
5. Cork discs, e.g., catalog number SKU: 63305 with 20 mm diameter and 3 mm thickness, Electron Microscopy Sciences.
6. Personal protective equipment including lab coat, eye protection glasses, and cryo-gloves.
7. Long forceps for holding a sample cork.
8. Isopentane in a metal or glass beaker with wire.
9. Liquid nitrogen in a cylindrical Dewar flask.
10. Dry ice.

2.5 Cryosectioning of the Muscle Samples

1. Antistripping coated slide glasses, e.g., MAS-coated slide glasses (Matsunami Glass Ind., Ltd.).
2. Fine brushes.
3. Optimal cutting temperature (O.C.T.) compound.
4. Rapid freezing spray to bring to ~ −60 °C.
5. Cryostat set at −25 °C.

2.6 Fluorescence in Situ Hybridization (FISH)

Reagents used in this method should be confirmed as "RNase-free" with the product information. Otherwise, DEPC treatment is recommended for some of them, as necessary. Labware should be treated with a dry heat sterilizer or RNase decontamination solution before use.

1. Fluorescent hybridization probe targeting CUG repeat RNA, e.g., $(CAG)_{5-7}$ peptide nucleic acid (PNA) probe conjugated to Cy3 via an O-linker at the N-terminal of the probe (*see* **Note 4**).
2. FITC-conjugated wheat germ agglutinin, 2 mg/mL in PBS(−) (e.g., catalog no. L4895, Merck). Make aliquots and store below −20 °C until use.
3. Antifade reagent with a fluorescent nucleic acid dye, e.g., Vectashield Hard Set Mounting Medium with DAPI (Vectashield) or TO-PRO-3.
4. Coplin jars for exclusive use for RNA experiments.
5. Moisture chamber.
6. Laboratory ovens.
7. 4% paraformaldehyde, purified prill.
8. Pre- and post-hybridization solutions containing 30% formamide in 40 mL each: Add 12 mL of formamide and 4 mL of 20× SSC into 24 mL of RNase-free water in a fresh 50 mL conical tube. Prepare just before use and keep at room temperature.
9. Hybridization solution in 600 μL: Into 312.8 μL of RNase-free water, add 200 μL of formamide, 60 μL of 20× SSC, 12 μL of 10 mg/mL bovine serum albumin, 4.2 μL of 10 mg/mL yeast transfer RNA, 6 μL of 200 mM ribonucleoside–vanadyl complex, and 5 μL of 10 μM PNA probe (*see* **Note 5**). Prepare the solution just before use.
10. Coverslips or parafilm to cover tissue sections for hybridization.
11. Permeabilization solution: 0.5% Triton X-100 in PBS(−). Store at room temperature.
12. 1× SSC, 40 mL in a fresh 50 mL conical tube.
13. Wash solution: 0.1% Tween 20 in PBS(−). Store at room temperature.
14. Nail polish.
15. Confocal fluorescence microscope or a similar one such as a BZ-X710 All-in-One fluorescence microscope (Keyence).

3 Methods

3.1 Genotyping of HSA^LR Mice

1. Administer an anesthetic to a mouse in the induction box of an anesthesia machine using 2–3% isoflurane (*see* **Note 6**).
2. Collect a small circle of the peripheral ear with a disinfected ear punch and place it in a 1.7 mL tube. The small tissue can be stored at −80 °C until use.
3. Extract genomic DNA using the DNeasy Blood & Tissue Kit according to the manufacturer's instructions.
4. Using a spectrophotometer, measure and adjust the extracted DNA concentration to 10–20 ng/μL.
5. Prepare PCR reaction mix in 50 μL including 0.3 μM primers and 50–100 ng of genomic DNA.
6. Conduct PCR with a step-down cycle program to prevent a smear in amplification. A sample program for the KOD FX Neo kit is shown in Table 2.
7. Electrophorese PCR products on a 2% agarose gel with a DNA stain at 100 mV for 30 min or more to make the CTG repeat lengths clearer between mouse samples.
8. Visualize the PCR bands with a UV transilluminator. An example is shown in Fig. 1.

Table 2
A sample step-down cycling program of genotyping PCR to detect CTG repeat DNA in *HSA*^LR mice, using a DNA polymerase to amplify a GC-rich region, KOD FX Neo

Steps	°C	Time	Cycle no.
Initial denaturation	94	2 m	
Denature	98	10 s	5
Extension	74	30 s	
Denature	98	10 s	5
Extension	72	30 s	
Denature	98	10 s	5
Extension	70	30 s	
Denaturation	98	10 s	30
Annealing	62	25 s	
Extension	68	30 s	
Cooling	4	5 m	
Hold	15	–	

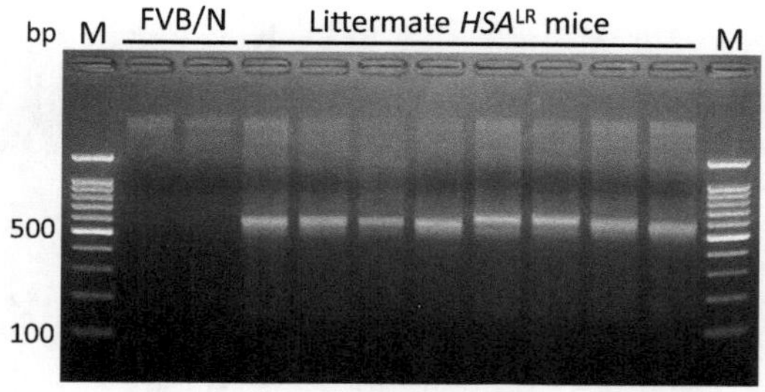

Fig. 1 A sample result for genotyping PCR of *HSA*^{LR} mice using a specific primer set to amplify the CTG repeat region in the transgene, human *ACTA1*. No bands are found in wild-type FVB/N mice. *HSA*^{LR} mice have various CTG repeat lengths even in littermates, as seen in different PCR amplicon sizes on 2% agarose gel prestained with ethidium bromide. The result may be considered for selecting mice used for breeding and experiments. M, 100 bp ladder

9. Process PCR bands with a DNA purification kit according to the manufacturer's instructions.
10. For sequencing to confirm the CTG repeat length, submit the purified PCR products and the specific primers to a company, e.g., Azenta Life Sciences (*see* **Note 7**).

3.2 Needle EMG for Hind Limb Muscles

1. Mat an electromagnetic wave-shielding sheet for EMG on a laboratory bench. An anesthesia system should not be on the sheet, so the system can be covered with the sheet (*see* **Note 8**). Place a supplemental heating source, such as a disposable pocket warmer or thermostatic heating plate, under the sheet to prevent a decrease in the body temperature of the mouse.
2. Before mouse preparation, connect concentric and monopolar (ground) needle electrodes to an EMG device and set the parameter to detect action potentials following a needle insertion according to the manufacturer's instructions (*see* **Note 9**).
3. Place a mouse in the anesthesia chamber of the machine. Use 4–5% isoflurane for the induction of anesthesia in a mouse and ensure the loss of the righting reflex and the reduction of breathing rates to approximately half. Then, switch the isoflurane concentration to 2–3% for maintenance (*see* **Note 10**).
4. Quickly take the mouse out of the chamber and set a nose cone of the anesthesia machine to a mouse.
5. Lay a mouse face-up. Extending a hind limb, shave or remove the hairs through a wide area covering measured muscles, such as the tibialis anterior, gastrocnemius, and quadriceps, of interest.
6. After changing the mouse's posture to face-down, disinfect the back of the neck with 70% ethanol and then insert the ground electrode under the neck skin.

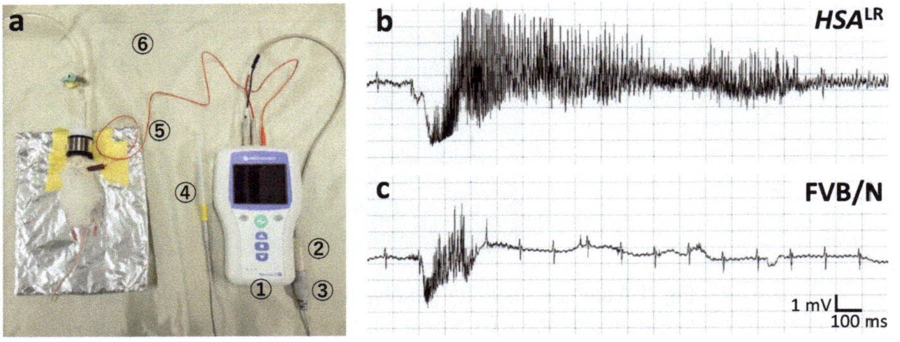

Fig. 2 Representative EMG result of the prolonged action potential duration (myotonic discharge) in an *HSA*^{LR} mouse under isoflurane anesthesia. (**a**) Example materials for a potable EMG device: (1) Neuropack n1 (Nihon Kohden), (2) female connector, (3) male connector, (4) concentric needle electrode connected to the male connector, (5) ground electrode, (6) electromagnetic wave-shielding sheet. (**b**) An example of a myotonic discharge as seen in the gastrocnemius muscle of an *HSA*^{LR} mouse. (**c**) An example EMG waveform of a typical short burst due to needle-insertional activity in the gastrocnemius muscle of a wild-type FVB/N mouse

7. Turn the mouse onto the side and start recording on the EMG device. Extending a hind limb with a nondominant hand, insert a concentric needle electrode subcutaneously into a hind limb muscle. Keeping the needle under the skin, confirm the stability of the EMG waveform shown in the device. Be careful not to insert the needle tip into the muscle at this point.
8. Keeping under the skin, gently insert the needle into a muscle of interest at an around 30–45 degree angle. In *HSA*^{LR} mice, the needle insertion represents a prolonged action potential duration and a characteristic "dive bomber" sound, which are hallmarks of myotonic discharge detected in myotonia due to the CTG repeat. Example images are shown in Fig. 2.
9. From the same position at the skin, approximately three insertions at different muscle positions are acceptable by moving the needle tip under the skin. After three muscle insertions from the same skin position, remove the needle from the skin and re-insert it from a different skin position for additional muscle measurements (*see* **Note 11**). Conduct a total of 10–15 insertions per muscle for the analysis.
10. Myotonic discharges recorded in exclusive software are evaluated as a percentage of the total number of insertions or as a grading score on a 4-point scale: 0, no myotonia; 1, occasional myotonic discharge in ≤50% of needle insertions; 2, myotonic discharge in >50% of insertions; and 3, myotonic discharge with nearly all insertions.
11. After the measurement, leave the mouse in a recovery cage with a supplementary heat source, e.g., a disposable pocket warmer wrapped in a paper towel, until it has retrieved consciousness and held sternal recumbency.

3.3 Preparation of Frozen Muscle Blocks

1. Euthanize a mouse by cervical dislocation under anesthesia with isoflurane according to an Institutional Animal Care and Use Committee-approved method.
2. Immediately harvest an entire muscle tissue of interest and vertically mount it onto tragacanth gum on a cork disc using fine forceps. The insertion (distal end) of muscles should be directed into the gum.
3. Holding the tissue-embedded cork disc with long forceps, soak and gently shake it into isopentane thoroughly cooled in liquid nitrogen for at least 1 min. During the procedure, maintain a higher level of liquid nitrogen in a cylindrical Dewar flask than that of isopentane in a beaker.
4. Place the snap-frozen tissue block on dry ice until isopentane on the block evaporates and then store them at −80 °C until use.

3.4 FISH to Detect CUG Repeat RNA on Skeletal Muscle Sections

Cryosections, pre- and post-hybridization solutions, and a hybridization solution should be prepared just before use to keep them fresh for RNA.

1. After cooling a cryostat chamber containing a microtome blade and fine brushes at −25 °C, transport the frozen muscle tissue blocks with dry ice and place the tissues into the chamber to allow them to equilibrate to the cryostat temperature for 10–30 min.
2. Prepare frozen skeletal muscle sections 7–10 μm thick on antistripping-coated slide glasses (*see* **Note 12**). The sections should be made from at least three levels of muscle positions at proper intervals, e.g., 100 or 200 μm intervals.
3. Dry the sections at room temperature for 30 min.
4. Fix the sections with 4% paraformaldehyde (PFA) in PBS(−) for 15 min at room temperature.
5. Transfer the slides into a Coplin jar and immerse them in PBS (−) for 2 min five times on an orbital rotator.
6. Incubate the slides in PBS(−) containing 0.5% Triton X-100 for 5 min at room temperature with shaking.
7. Immerse and incubate the slides in the prehybridization solution at room temperature for 10 min with shaking.
8. Put the hybridization solution involving the Cy3-labeled (CAG)n PNA probe onto the slides and cover it with a coverslip or parafilm.
9. Place the probe solution-applied slides in a moisture chamber shaded from light, such as using tinfoil, and incubate it in an oven at 37 °C overnight.

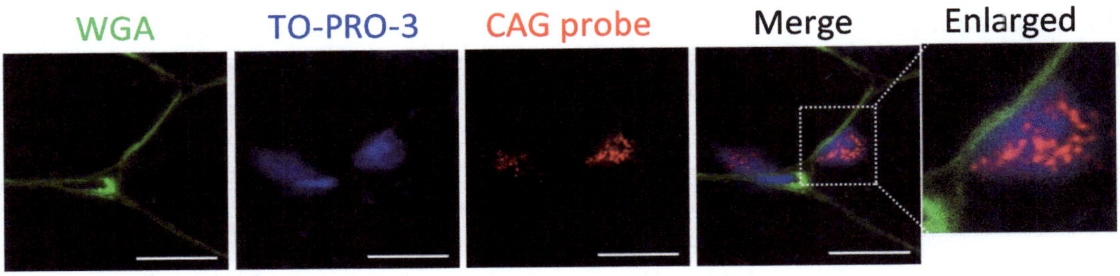

Fig. 3 An example of a FISH result on a cryosection of the gastrocnemius muscle in an *HSA*^{LR} mouse. Muscle fibers are outlined in green with FITC-labeled wheat germ agglutinin (WGA). Nuclei are highlighted in blue with TO-PRO-3 stain. Myonuclear RNA foci derived from expanded CUG repeat RNA, as detected by a Cy3-conjugated $(CAG)_7$ PNA probe, are shown in red. Bar, 10 μm

10. Prewarm the post-hybridization solution in a Coplin jar at 42 °C in an oven.
11. Immerse and incubate the slides in the post-hybridization solution in an oven at 42 °C for 30 min.
12. Wash the slides with 1× SSC in a Coplin jar for 30 min at room temperature with shaking.
13. Wash the slides with PBS(−) containing 0.1% Tween 20 for 10 min at room temperature on an orbital rotator twice.
14. Apply FITC-conjugated wheat germ agglutinin at 20 μg/mL in PBS(−) to the slides and incubate them for 10 min at room temperature.
15. Wash the slides with PBS(−) containing 0.1% Tween 20 for 5 min at room temperature on an orbital rotator twice.
16. Put an antifade reagent with a nuclear counterstain, e.g., Vectashield Hard Set Mounting Medium with DAPI, on the sections according to the manufacturer's instructions (*see* **Note 13**). Cover it with a coverslip and remove the excess solution using a laboratory wipe.
17. Fix the mounting medium in the slides at 4 °C for several hours or overnight shaded from light. Then, seal the coverslip with nail polish and observe the fluorescent signals of RNA foci. An example result is shown in Fig. 3.

4 Notes

1. After being introduced into an animal facility for research, mice are maintained and bred under a standard 12 h light–dark cycle with free access to food and water, according to a method approved by an Institutional Animal Care and Use Committee.

2. The length of CTG repeat in *HSA*^{LR} mice varies even between littermates. Genotyping PCR results may serve to see the potential effects of the CTG repeat length on other application results of mice.
3. Tragacanth gum powder is mixed and tempered with an equal volume of distilled water using a spatula. The softness of the gum can be altered by the amount of water. Once the gum becomes semisoft and sticky, it is stuffed into a 5 mL syringe using a spatula, which can be stored at −20 °C until use.
4. PNA probes are dissolved to a concentration of 10–100 μM in formamide, and the aliquots are stored below −20 °C, protected from light. Before making the aliquots, the PNA probe stock can be heated at 55 °C for 5 min to be fully dissolved because purine-rich sequences may have low solubility.
5. Heat the PNA probe aliquot at 85 °C for 5 min before using it, if necessary.
6. Place a mouse in the induction box of an anesthesia machine and ensure that the mouse has lost the righting reflex and the breathing rate has been reduced to ~50%.
7. To read DNA sequences with a high GC content and a trinucleotide repeat, select an appropriate sequencing method in consultation with the company being used.
8. An electromagnetic wave-shielding sheet for EMG can help prevent an EMG device from detecting unintended electromagnetic waves from an anesthesia machine or surrounding environment.
9. In an example of a portable EMG device, Neuropack n1 (MEM-8301, Nihon Kohden), select 50 ms/division for imaging speed, 1 mV/division for display sensitivity, and no alternating current (AC) filter if there are no AC interferences in an environment where EMG is performed.
10. Because anesthetic sensitivity varies depending on the conditions of mice, such as age, body weight, and treatment, adjust the anesthetic percentage for maintenance.
11. Repetitive insertions into a muscle from the same subcutaneous position reduce the reactivity of the muscle to the needle insertion, which prevents the accurate measurement of myotonic discharge. Removing the needle from the skin, re-insert it from a different position of the skin so that the needle can approach different muscle areas.
12. Before starting cryosectioning, ensure there are no cracks in a tissue block. Cracks in the gum stage can cause difficulty in making even cryosections. To fix the cracks, quickly rub an appropriate amount of the O.C.T. compound on the cracks and immediately fix it with a freezing spray.

13. In the case of using a separate nuclear counterstain, such as TO-PRO-3, after counterstaining nuclei with the agent according to the manufacturer's instructions, wash the slides with PBS(−) containing 0.1% Tween 20 for 10 min three times at room temperature on an orbital rotator. Following the mounting of an antifade reagent on the sections, gently place a coverslip over them and close it with nail polish.

Acknowledgments

The authors thank Dr. Masato Kitagawa (Nihon University Animal Medical Center) and Yume Shinohara for assistance of EMG. This work was partially supported by the Japan Agency for Medical Research and Development (grant number 22ek0109496h0002 to Y.A.), the Intramural Research Grant (2–7, 5–6) for Neurological and Psychiatric Disorders of National Center of Neurology and Psychiatry (NCNP), and JSPS KAKENHI (grant number 21 K06976 to YE).

Conflict of Interest The authors have no conflicts of interest to declare that are relevant to the content of this chapter.

References

1. Johnson NE, Butterfield RJ, Mayne K, Newcomb T, Imburgia C, Dunn D, Duval B, Feldkamp ML, Weiss RB (2021) Population-based prevalence of myotonic dystrophy type 1 using genetic analysis of statewide blood screening program. Neurology 96(7):e1045–e1053. https://doi.org/10.1212/wnl.0000000000011425
2. Thornton CA (2014) Myotonic dystrophy. Neurol Clin 32(3):705–719, viii. https://doi.org/10.1016/j.ncl.2014.04.011
3. Pascual-Gilabert M, Artero R, López-Castel A (2023) The myotonic dystrophy type 1 drug development pipeline: 2022 edition. Drug Discov Today 28:103489. https://doi.org/10.1016/j.drudis.2023.103489
4. Nakamori M, Nakatani D, Sato T, Hasuike Y, Kon S, Saito T, Nakamura H, Takahashi MP, Hida E, Komaki H, Matsumura T, Takada H, Mochizuki H (2024) Erythromycin for myotonic dystrophy type 1: a multicentre, randomised, double-blind, placebo-controlled, phase 2 trial. EClinicalMedicine 67:102390. https://doi.org/10.1016/j.eclinm.2023.102390
5. Nakamori M (2024) Expanded-repeat-RNA-mediated disease mechanisms in myotonic dystrophy. Neurol Clin Neurosci 12(1):16–23. https://doi.org/10.1111/ncn3.12687
6. Gomes-Pereira M, Cooper TA, Gourdon G (2011) Myotonic dystrophy mouse models: towards rational therapy development. Trends Mol Med 17(9):506–517. https://doi.org/10.1016/j.molmed.2011.05.004
7. Mankodi A, Logigian E, Callahan L, McClain C, White R, Henderson D, Krym M, Thornton CA (2000) Myotonic dystrophy in transgenic mice expressing an expanded CUG repeat. Science 289(5485):1769–1773
8. Braz SO, Acquaire J, Gourdon G, Gomes-Pereira M (2018) Of mice and men: advances in the understanding of neuromuscular aspects of myotonic dystrophy. Front Neurol 9:519. https://doi.org/10.3389/fneur.2018.00519
9. Wheeler TM, Sobczak K, Lueck JD, Osborne RJ, Lin X, Dirksen RT, Thornton CA (2009) Reversal of RNA dominance by displacement of protein sequestered on triplet repeat RNA. Science 325(5938):336–339. https://doi.org/10.1126/science.1173110

10. Nakamori M, Taylor K, Mochizuki H, Sobczak K, Takahashi MP (2016) Oral administration of erythromycin decreases RNA toxicity in myotonic dystrophy. Ann Clin Transl Neurol 3(1):42–54. https://doi.org/10.1002/acn3.271

11. Wheeler TM, Leger AJ, Pandey SK, MacLeod AR, Nakamori M, Cheng SH, Wentworth BM, Bennett CF, Thornton CA (2012) Targeting nuclear RNA for in vivo correction of myotonic dystrophy. Nature 488(7409):111–115. https://doi.org/10.1038/nature11362

Chapter 12

Utilization of Induced Pluripotent Stem Cell-Derived Neurons to Investigate the Splice-Modification Efficacy of Splice-Switching Drug Candidates

Yukiko Ueta, Hiroaki Ohara, Tomonari Awaya, and Masatoshi Hagiwara

Abstract

Modification of RNA splicing represents a promising therapeutic strategy for treating genetic diseases. Several splice-switching drugs have already been approved, notably for spinal muscular atrophy and specific genotypes of Duchenne muscular dystrophy. While peripheral blood mononuclear cells and fibroblasts are commonly used to evaluate the efficacy of these drug candidates in patient-derived samples, such approaches may lead to misleading conclusions due to the cell type-specific nature of RNA splicing.

Induced pluripotent stem cells (iPSCs), which can be differentiated into various somatic cell types, offer a valuable tool to overcome this limitation. Once efficient differentiation protocols are established, iPSCs can give rise to neurons, retinal cells, cardiomyocytes, and other hard-to-obtain cell types. In this chapter, we describe a protocol for evaluating the splice-modification efficacy of drug candidates in iPSC-derived neurons, using the removal of a poison exon in *SCN1A* gene, associated with Dravet syndrome as a model.

Key words RNA splicing, Splice-switching oligonucleotides (SSOs), Induced pluripotent stem cells (iPSCs), Functional validation of splicing correction, Antisense oligonucleotides, Small-molecule compounds

1 Introduction

Alternative splicing is a fundamental biological process that expands the diversity of the transcriptome and proteome. Its dysregulation is increasingly recognized as a contributor to various genetic diseases. To correct aberrant splicing patterns, splice-switching oligonucleotides (SSOs) and other splice-modifying drugs have been developed, some of which have already reached clinical use. Notable examples include nusinersen [1] and risdiplam [2] for spinal muscular atrophy, as well as several exon-skipping drugs approved for specific mutations in Duchenne muscular dystrophy [3]. These successes have accelerated efforts to develop similar therapies targeting other splicing-related disorders.

Yoshitsugu Aoki and Norio Motohashi (eds.), *Splice Switching: Methods and Protocols*, Methods in Molecular Biology, vol. 2963, https://doi.org/10.1007/978-1-0716-4738-7_12,

To evaluate the efficacy of splice-modifying drug candidates, researchers often utilize patient-derived cell lines such as peripheral blood mononuclear cells (PBMCs), fibroblasts, or Epstein–Barr virus-transformed B-lymphoblastoid cells. While these models are readily accessible and easy to handle, they may not reflect the splicing environment of the disease-relevant cell type, as splicing regulation is often cell-type-specific. Consequently, reliance on such surrogate models can lead to either underestimation or overestimation of drug efficacy.

Induced pluripotent stem cells (iPSCs) provide a powerful solution to this limitation. iPSCs can be differentiated into disease-relevant cell types, including neurons, cardiomyocytes, and retinal cells, making them an attractive platform for disease modeling and therapeutic screening [4]. In this chapter, we present a protocol to evaluate the efficacy of splice-switching drug candidates in neurons differentiated from patient-derived iPSCs. This approach allows for the quantitative assessment of exon inclusion or skipping in the appropriate cellular context and offers a versatile tool for preclinical validation of splicing-modifying therapies.

In this chapter, we demonstrate the utility of iPSC-derived neurons in drug discovery, using Dravet syndrome (DS) as a representative example where splicing modulation offers a promising therapeutic strategy. DS is a severe epileptic encephalopathy resulting from haploinsufficiency of the *SCN1A* gene due to a loss-of-function mutation or deletion in one allele [5]. Recent studies have shown that the remaining wild-type allele retains the potential to produce sufficient levels of functional sodium channels [6]. Notably, the wild-type *SCN1A* gene harbors a 64-nucleotide poison exon, commonly referred to as exon 20 N, located between canonical exons 20 and 21. Inclusion of exon 20 N introduces a frameshift and a premature termination codon, leading to nonsense-mediated mRNA decay (NMD), which may further reduce the levels of functional transcripts. Therefore, suppressing the inclusion of exon 20 N may enhance the production of functional isoforms and thereby partially compensate for the haploinsufficiency. Using patient-derived iPSC neurons, we present a protocol to evaluate the efficacy of splice-switching drug candidates in this disease-relevant cellular model.

A number of small molecules have been reported as RNA splicing modifiers. Here, we evaluated the effects of five splicing-modifying compounds—RECTAS (10 μM, a CLK activator) [7, 8], SRPIN340 (30 μM, an SRPK inhibitor) [9], and two CLK inhibitors, TG003 (30 μM) [10], CaNDY (5 μM) [11], and ALG2 (10 μM, a DYRK inhibitor) [12]—on *SCN1A* exon 20 N splicing in iPSC-derived neurons. All of these compounds are original small molecules previously developed by our group. In addition, 100 μg/mL cycloheximide (CHX) was used to assess the involvement of nonsense-mediated mRNA decay (NMD). The iPSC-derived neurons exhibited several aberrant isoforms, as well as the normal productive isoform, as shown in Fig. 1. CHX treatment revealed that the exon 20 N-containing isoforms were subjected to NMD (lanes 1 and 2). The CLK inhibitors restored the normal productive isoform, whereas RECTAS and SRPIN340 did not (lanes 3–7). The neuronal differentiation protocol is summarized in Fig. 2.

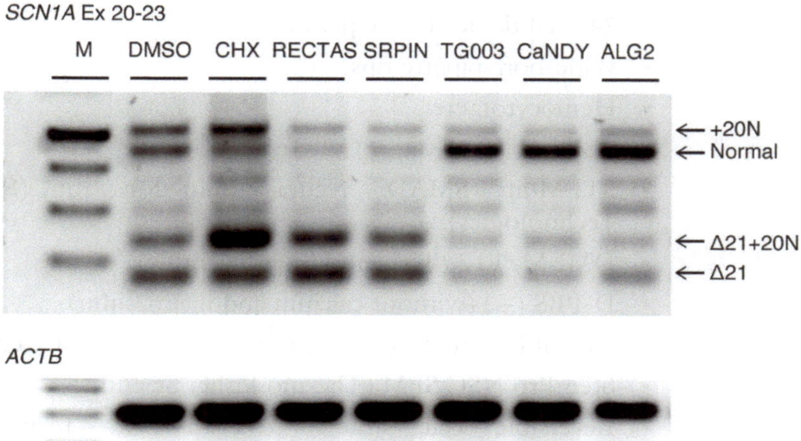

Fig. 1 Changes in the splicing pattern of *SCN1A* transcripts in iPSC-derived neurons treated with splice-modifying compounds. Day 40 neurons were dissociated and reseeded at a density of 5×10^4 cells/cm^2. Fourteen days after reseeding, the neurons were treated with 0.1% DMSO, CHX (100 μg/mL), RECTAS (10 μM), SRPIN340 (30 μM), TG003 (30 μM), CaNDY (5 μM), and ALG2 (10 μM). Cells were harvested 24 h after treatment, and RT-PCR was performed as described above. The results showed the presence of several aberrant isoforms (519 bp: +20 N; 237 bp: Δ21 + 20 N; 173 bp: Δ21) along with the normal productice isoform (455 bp). Additional bands may represent heteroduplexes, as RNA-seq did not reveal other distinct isoforms (data not shown). *M* molecular weight marker

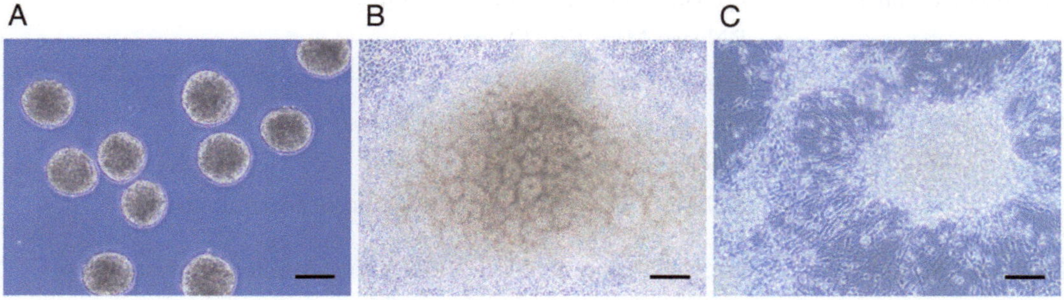

Fig. 2 Stepwise differentiation of iPSCs into cortical neurons. (**a**) Formation of uniform embryoid bodies (EBs) in AggreWell 800 plates on day 5. EBs exhibit a compact, spherical morphology. (**b**) Neural rosettes on day 12. Radially aligned neuroepithelial cells form characteristic rosette-like structures. (**c**) Neurons on day 47. Cells were dissociated and reseeded on day 33 and cultured for additional 14 days. Neuronal cells aggregate and extend the axons to form a neural network. Scale bar = 200 μm

2 Materials

2.1 iPSC Culture and Differentiation

- StemFit AK02N (Ajinomoto).
- iMatrix-511 silk (Matrixome).
- Y-27632 (ROCK inhibitor).
- Matrigel (Corning).
- Six-well tissue culture plates.
- 24-well tissue culture plates.
- Wide-bore pipette tips.
- Hemocytometer.
- Accutase (Innovative Cell Technologies).
- STEMdiff SMADi Neural Induction Kit (STEMCELL Technologies).
- DMEM/F12 (Thermo Fisher Scientific).
- D-PBS (−) (without calcium and magnesium).
- Neural Rosette Selection Reagent (STEMCELL Technologies).
- StemPro NSC SFM (Thermo Fisher Scientific).
- Reversible cell strainer (STEMCELL Technologies or equivalent).

2.2 Small-Molecule Treatment (See Note 1)

- RECTAS (CLK activator).
- SRPIN340 (SRPK inhibitor).
- TG003 (CLK inhibitor).
- CaNDY (CLK inhibitor).
- ALG2 (DYRK inhibitor).
- Dimethyl sulfoxide (DMSO).

2.3 RNA Extraction

- QIAshredder spin columns (Qiagen).
- RNeasy Mini Kit (Qiagen).
- β-Mercaptoethanol.
- 70% ethanol (prepared with RNase-free water).
- Nuclease-free water.
- RNase-free 1.5 mL and 2 mL microcentrifuge tubes.

2.4 cDNA Synthesis

- High-Capacity RNA-to-cDNA Kit (Applied Biosystems).
- Nuclease-free water.
- Thermal cycler.

2.5 PCR and Electrophoresis

- AmpliTaq Gold 360 Master Mix (Thermo Fisher Scientific).
- Forward and reverse primers (10 μM stocks).
- PCR tubes or plates.
- Agarose.
- TAE or TBE buffer.
- DNA ladder.
- Gel electrophoresis system.
- Gel imaging system.

3 Methods

3.1 Maintenance of Undifferentiated iPSCs

Culture iPSCs with StemFitAK02N and iMatrix-511 silk according to the manufacturer's protocol. We use a six-well plate format for daily maintenance with seeding density at 1300 cells/well (*see* **Note 2**).

3.2 Differentiation of iPSCs into Cortical Neurons

Our current protocol has been established by combining reagents from multiple suppliers. It is applicable to a broad range of iPSC lines cultured with StemFit AK02N and iMatrix-511. Depending on the neural induction efficiency of the iPSC line, the Rosette Selection step can sometimes be omitted (*see* **Note 3**). Neurons cultured for 2–4 weeks in StemPro NSC SFM, typically from day 33 (or day 26) to day 47 (or day 40), are generally suitable for downstream applications. Single-cell dissociation and reseeding are recommended to obtain homogeneous samples. For in vitro drug testing, 24-well plates are appropriate for collecting RNA and protein samples.

Day 0: Formation of Embryoid Bodies (EBs)

1. Rinse each well of AggreWell 800, 24-well plate with 1.0 mL of DMEM/F12. Add 1.0 mL STEMdiff NIM + SMADi. Centrifuge the plate at 3000 × g for 5 min so that bubbles are removed from microwells. Place the plate in a 37 °C, 5% CO_2 environment.
2. Rinse iPSCs with 2.0 mL D-PBS. Add 500 μL Accutase and put the plate in a 37 °C, 5% CO2 environment for 3–5 min.
3. Add 4.5 mL of DMEM/F12 to the plate and collect cells to a 15 mL tube.
4. Count the number of cells and transfer 3×10^6 cells to another 15 mL tube. Centrifuge at 300 × g for 5 min at room temperature. While centrifuging, prepare a mixture of 1.0 mL STEMdiff NIM + SMADi and 2.0 μL of 10 mM Y-27632.

5. Suspend the cell pellet in 1 mL of the prepared STEMdiff NIM + SMADi +Y-27632 and place it in an AggreWell 800 plate. Mix by pipetting and centrifuge at 100 × g for 3 min at room temperature for the cells to settle down into microwells.
6. Culture the cells in a 37 °C, 5% CO_2 environment. Replace 3/4 medium every other day with STEMdiff NIM + SMADi without Y-27632. Be careful not to let EBs float when replacing the medium.

Day 5: Attach EBs to a Matrigel-Coated Plate

1. Prepare a Matrigel-coated plate in advance. Briefly, dilute Matrigel matrix in ice-cooled DMEM/F12 according to manufacturer's instructions. Use "dilution factor" (indicated on the data sheet) μL of Matrigel matrix for 25 mL of DMEM/F12. Keep the plate at least 30 min at room temperature. Do not remove the Matrigel solution until use.
2. Check the morphology of the EBs in the AggreWell 800 plate under an inverted microscope. Proceed to the next step only if the EBs are densely aggregated and have a smooth, spherical appearance (Fig. 2a).
3. Place a reversible cell strainer on a 15 mL conical tube.
4. Gently add DMEM/F12 to the well to let EBs float. Collect the EBs onto the reversible cell strainer using a wide bore tip. Repeat this operation so that you will collect as many EBs as possible. Do not treat EBs harshly.
5. Aspirate Matrigel solution from the well of the Matrigel-coated six-well cell culture plate. Flip the reversible cell strainer over the well and add 2.0 mL STEMdiff NIM + SMADi (Y-27632 not required) to the strainer to seed EBs into the well.
6. Gently move the plate in random directions to distribute EBs evenly. Culture the EBs in a 37 °C, 5% CO_2 environment. Do not move the plate 24 h to allow the EBs to attach to the plate. Replace the media every other day.

Day 12: Select Neural Rosettes

1. Prepare a Matrigel-coated six-well plate in advance, as described above. Keep the plate at least 30 min at room temperature. Do not remove the Matrigel solution until use.
2. Check the neural rosette structure under an inverted microscope. Proceed to the next step only if the rosettes are well-formed, with radially arranged cells and a clear organized structure (Fig. 2b).
3. Rinse the culture surface with DMEM/F12. Add 1.0 mL of Neutral Rosette Selection Reagent and incubate in a 37 °C, 5% CO2 environment for 30 min to 1 h.

4. Spray DMEM/F12 with a 1000 μL micropipette to the surface of the rosette structures. Repeat until the most rosette structures detach. Be careful not to break the rosettes into single cells with excessive pipetting, which may reduce the cell viability. Also do not collect the cells surrounding the rosette much, since they mostly consist of neural crest cells and non-neuronal lineage cells.
5. Collect the detached rosettes and cells into a 15 mL conical tube, centrifuge at 350× *g* for 5 min at room temperature, and aspirate the supernatant.
6. Resuspend cells in 2 mL of STEMdiff NIM (without SMADi) and seed into one well of a Matrigel-coated six-well plate. Culture the cells in a 37 °C, 5% CO_2 environment. Replace the media every other day.

Day 19 (or Day 12): Differentiation into Neurons

1. Change the medium from STEMdiff NIM to StemPro NSC SFM. Aspirate the media and add 2.0 mL StemPro NSC SFM.
2. Replace half of the medium three times per week. The cells will differentiate into neurons and extend their axons (*see* **Note 4**).

Day 33–47 (or Day 26–40): Preparation of the Neurons for In Vitro Drug Testing

1. Prepare a Matrigel-coated 24-well plate in advance, as described above. Keep the plate at least 30 min at room temperature. Do not remove the Matrigel solution until use.
2. Rinse the cells with 2 mL of D-PBS(−). Add 1 mL Accutase and incubate in a 37 °C, 5% CO2 environment for 10–15 min.
3. Gently pipette to dissociate the cells into a single-cell suspension. Dilute the suspension with 9.0 mL of DMEM/F12 and transfer to a 15 mL centrifuge tube.
4. Count the cell numbers using a hemocytometer.
5. Centrifuge at 350× *g* for 5 min at room temperature and then aspirate the supernatant.
6. Resuspend the cell pellet in StemPro NSC SFM to prepare a cell suspension at 2–6 × 10^5 cells/mL.
7. Seed 500 μL of the suspension into each well of the 24-well plate, resulting in a final density of 1–3 × 10^5 cells per well (5 × 10^4–1.5 × 10^5 cells/cm^2) (Fig. 2c; *see* **Note 5**).

3.3 Compound Treatment

To simplify the experimental procedure, we first prepared 1000× stock solutions of each compound in DMSO, resulting in a final DMSO concentration of 0.1% in all treatment conditions.

1. Prepare StemPro NSC SFM supplemented with each small molecule. In this protocol, RECTAS, SRPIN340, TG003,

CaNDY, and ALG2 were diluted in StemPro NSC SFM to final concentrations of 10 μM, 30 μM, 30 μM, 5 μM, and 10 μM, respectively. Be sure to include DMSO treatment as a control.

2. Aspirate the culture medium from a 24-well culture plate by a micropipette and perform a full medium change with the prepared medium.
3. Return to the incubator and culture the cells in a 37 °C, 5% CO_2 environment.

3.4 RNA Extraction (See Note 6)

1. Rinse adherent cells once with 500 μL of D-PBS (−).
2. Add 350 μL of buffer RLT supplemented with 1% β-mercaptoethanol directly to the cells.
3. Lyse the cells by pipetting up and down gently. Avoid generating bubbles.
4. Transfer the lysate to a QIAshredder spin column placed in a 2 mL collection tube.
5. Centrifuge at a maximum speed (≥13,000 × g) for 2 min at room temperature.
6. Transfer the flow-through to a new RNase-free microcentrifuge tube.
7. Add 350 μL of 70% ethanol to the lysate and mix thoroughly by pipetting.
8. Transfer up to 700 μL of the mixture to an RNeasy spin column placed in a new 2 mL collection tube.
9. Centrifuge at ≥8000 × g for 15 s at room temperature. Discard the flow-through.
10. Add 700 μL of buffer RW1 to the column. Centrifuge at ≥8000 × g for 15 s. Discard the flow-through.
11. Add 500 μL of buffer RPE to the column. Centrifuge at ≥8000 × g for 15 s. Discard the flow-through.
12. Add another 500 μL of buffer RPE to the column. Centrifuge at ≥8000 × g for 2 min.
13. Place the column in a new collection tube and centrifuge at full speed for 1 min to remove residual ethanol.
14. Transfer the RNeasy spin column to a new 1.5 mL RNase-free microcentrifuge tube.
15. Add 30–50 μL of RNase-free water directly to the membrane.
16. Incubate at room temperature for 1–2 min.
17. Centrifuge at ≥8000 × g for 1 min to elute RNA.
18. Measure the RNA concentration and confirm that sufficient RNA has been obtained for RT-PCR. If not used immediately, store the RNA at −20 °C.

3.5 cDNA Synthesis and Agarose Gel Electrophoresis

1. According to the manufacturer's protocol for High-Capacity RNA-to-cDNA Kit, thaw the 2× RT Buffer Mix and 20× RT Enzyme Mix on ice. Mix well and briefly centrifuge. Keep on ice until use.
2. Prepare the reverse transcription reaction mix on ice according to the table below:

Component	Volume per 20 μL reaction
2× RT buffer mix	10.0 μL
20× RT enzyme mix	1.0 μL
Nuclease-free water	x μL (q.s. to 20 μL)
RNA template	y μL (up to 2 μg total RNA)
Total	**20.0 μL**

RNA template

3. Mix gently by pipetting and briefly centrifuge to collect contents.
4. Place the tubes into a thermal cycler and run the following program:
   - 37 °C for 60 min
   - 95 °C for 5 min
   - Hold at 4 °C.
5. Use the synthesized cDNA immediately or store at −20 °C until further use.

3.6 PCR and Agarose Gel Electrophoresis

1. Thaw all reagents on ice and mix well by gentle vortexing. Briefly spin down before use.
2. Prepare the PCR mix on ice according to the table below:

Comonent	Volume per 20 μL reaction
AmpliTaq gold 360 master mix (2×)	10.0 μL
Forward primer (10 μM)	0.5 μL
Reverse primer (10 μM)	0.5 μL
Total	**19.0 μL**

3. Dispense 19 μL of the reaction mix into each PCR tube. Add 1 μL of cDNA template per reaction.
4. Mix gently by pipetting and briefly centrifuge to collect contents.

5. Place the tubes in a thermal cycler and run the following program: 95 °C/10 min—(95 °C/30 sec—55 °C/30 sec—72 °C/60 sec) × indicated cycles—72 °C/5 min—4 °C/hold

   *PCR cycles were 35 cycles for *SCN1A* and 25 cycles for *ACTB*.

6. Run 5 μL of the PCR product on a 2% agarose gel with an appropriate DNA ladder (*see* **Note 7**).

4 Notes

1. In addition to the CLK inhibitors used in this protocol, various small molecules have been reported as RNA splicing modulators with distinct mechanisms of action. SF3B1 inhibitors such as *Spliceostatin A*, *FR901464*, *E7107*, and *H3B-8800* disrupt the U2 snRNP complex and are widely used to study the core spliceosome function. Related natural or synthetic compounds, including *Sudemycin D6* and *Herboxidiene*, share this mechanism but differ in the chemical structure and potency. In contrast, *Isoginkgetin* impairs spliceosome assembly at an earlier stage, independent of SF3B1. Additionally, certain dietary polyphenols—such as *epigallocatechin gallate (EGCG)*, *resveratrol*, and *α-tocopherol* (vitamin E)—have been reported to modulate alternative splicing indirectly, through effects on the expression or phosphorylation of splicing-regulatory proteins. These compounds provide a broad toolkit for investigating both mechanistic and therapeutic aspects of splicing regulation.
2. It is essential to prepare high-quality iPSCs in an undifferentiated state. The number of days between passages and the seeding density may need to be optimized depending on the specific iPSC line.
3. In certain iPSC lines, neural induction may proceed efficiently without the Rosette Selection step. In such cases, the overall differentiation period can be shortened by approximately 5–7 days (e.g., day 26 instead of day 33, as indicated in parentheses). However, omitting the rosette selection step may result in increased heterogeneity due to the presence of neural crest–derived or other non-neuronal cells.
4. The neurons can be maintained for several months to years, and long-term culture may promote neuronal maturation. However, prolonged culture often leads to cell aggregation, making subsequent dissociation difficult.
5. For short-term cultures (1–2 weeks), relatively high seeding densities (e.g., 1–3 × 10^5 cells/well) are acceptable and can

increase the yield of RNA or protein samples. However, for long-term culture, reducing the initial cell density is recommended. Higher densities often lead to excessive neuronal aggregation, resulting in spheroid-like structures and detachment of the cells from the plate over time.

6. For assessing splicing patterns, a 6-h treatment is generally sufficient. To evaluate the effects on protein expression, the incubation time can be extended to 24–72 h, depending on the half-life of the target protein.
7. When analyzing the splicing patterns by RT-PCR, followed by agarose gel electrophoresis, it is important to recognize that multiple transcript isoforms are amplified using the same primer set. This can introduce amplification bias due to the differences in amplicon length or GC content. Additionally, heteroduplex formation between similar PCR products during the cooling phase can result in aberrant bands with altered mobility. Interpretation should be made with caution. For more accurate quantification, complementary approaches such as qRT-PCR or RNA sequencing are recommended.

Conflict of Interest

The authors have no conflicts of interest to declare that are relevant to the content of this chapter.

References

1. Finkel RS, Mercuri E, Darras BT et al (2017) Nusinersen versus sham control in infantile-onset spinal muscular atrophy. N Engl J Med 377:1723–1732. https://doi.org/10.1056/NEJMoa1702752
2. Darras BT, Masson R, Mazurkiewicz-Bełdzińska M et al (2021) Risdiplam-treated infants with type 1 spinal muscular atrophy versus historical controls. N Engl J Med 385(5): 427–435. https://doi.org/10.1056/NEJMoa2102047
3. Matsuo M (2025) 30 years since the proposal of exon skipping therapy for Duchenne muscular dystrophy and the future of pseudoexon skipping. Int J Mol Sci 26(3):1303. https://doi.org/10.3390/ijms26031303
4. Takahashi K, Tanabe K, Ohnuki M et al (2007) Induction of pluripotent stem cells from adult human fibroblasts by defined factors. Cell 131: 861–872. https://doi.org/10.1016/j.cell.2007.11.019
5. Scheffer IE, Nabbout R (2019) SCN1A-related phenotypes: epilepsy and beyond. Epilepsia 60(S3):S17–S24. https://doi.org/10.1111/epi.16395
6. Han Z, Chen C, Christiansen A et al (2020) Antisense oligonucleotides increase Scn1a expression and reduce seizures and SUDEP incidence in a mouse model of Dravet syndrome. Sci Transl Med 12(558):eaaz6100. https://doi.org/10.1126/scitranslmed.aaz6100
7. Yoshida M, Kataoka N, Miyauchi K et al (2015) Rectifier of aberrant mRNA splicing recovers tRNA modification in familial dysautonomia. Proc Natl Acad Sci USA 112(9):2764–2769. https://doi.org/10.1073/pnas.1415525112
8. Ajiro M, Awaya T, Kim YJ et al (2021) Therapeutic manipulation of IKBKAP mis-splicing with a small molecule to cure familial dysautonomia. Nat Commun 12(1):4507. https://doi.org/10.1038/s41467-021-24705-5
9. Fukuhara T, Hosoya T, Shimizu S et al (2006) Utilization of host SR protein kinases and RNA-splicing machinery during viral

replication. Proc Natl Acad Sci 103:11329–11333. https://doi.org/10.1073/pnas.0604616103

10. Nishida A, Kataoka N, Takeshima Y et al (2011) Chemical treatment enhances skipping of a mutated exon in the dystrophin gene. Nat Commun 2:308. https://doi.org/10.1038/ncomms1306
11. Shibata S, Ajiro M, Hagiwara M (2020) Mechanism-based personalized medicine for cystic fibrosis by suppressing pseudo exon inclusion. Cell Chem Biol 27:1472–1482.e6. https://doi.org/10.1016/j.chembiol.2020.08.013
12. Nakano-Kobayashi A, Fukumoto A, Morizane A, Nguyen DT, Le TM, Hashida K, Hosoya T, Takahashi R, Takahashi J, Hori O, Hagiwara M (2020) Therapeutics potentiating microglial p21-Nrf2 axis can rescue neurodegeneration caused by neuroinflammation. Sci Adv 6(46):eabc1428. https://doi.org/10.1126/sciadv.abc1428

Chapter 13

Development of Electric Field Stimulation (EFS)-Mediated Skeletal Muscle Training with iPSCs

Tomoya Uchimura and Hidetoshi Sakurai

Abstract

Functional analysis of skeletal muscle in vitro requires maturation and stimuli to induce excitation–contraction coupling as well as a quantitative method to analyze contraction. Electric field stimulation (EFS) has been traditionally used as a method to excite skeletal muscle in vitro and in vivo. A combination of human-induced pluripotent stem cell (hiPS) technology and EFS will lead to recapitulation of functional phenotypes of skeletal muscle diseases. Here, we describe the combinational technology using hiPS technology and EFS to induce the maturation of skeletal muscle and analyze contractile performance in vitro.

Key words Induced pluripotent cells, Skeletal muscle, Electric field stimulation (EFS), Contraction, Gel culture

1 Introduction

Electric stimuli play an important physiological role in the regulation of skeletal muscle contraction as well as calcium signaling [1–7]. Electric field stimulation (EFS) has traditionally been used to mimic these stimuli in vitro. In general, skeletal muscle cells are stimulated by motor neuron leading to calcium mobilization and contraction [8]. To replicate this in vitro, hiPSC-derived skeletal muscle cells must be co-cultured with motor neurons, and a neuromuscular junction (NMJ) needs to be established. In the file of skeletal muscle, EFS can be used to stimulate skeletal muscle cells by inducing membrane depolarization and excitation–contraction coupling in the absence of motor neurons and NMJs. Alternatively, in the absence of motor neurons and NMJs, EFS can be used to directly stimulate hiPSC-derived skeletal muscle cells by inducing membrane depolarization and excitation–contraction coupling. Additionally, it is known that EFS can promote the maturation of skeletal muscle cells [1–7]. Thus, EFS enables researchers to mimic

Yoshitsugu Aoki and Norio Motohashi (eds.), *Splice Switching: Methods and Protocols*, Methods in Molecular Biology, vol. 2963,
https://doi.org/10.1007/978-1-0716-4738-7_13,

in vivo exercise in vitro for investigating the contractile performance of skeletal muscle cells. However, the utility of EFS for investigating hiPSC-derived skeletal muscle cells has been elusive. Here, we demonstrate that EFS can efficiently promote maturation of skeletal muscle cells differentiated from hiPSCs. Maturated skeletal muscles can be characterized by the presence of sarcomere structure as well as contractile performance. Contractile performance can be quantified using motion capture imaging system in conjunction with EFS. Time-dependent changes in contractile performance can be monitored by applying EFS while imaging with an inverted microscope, thereby quantifying how contractile performance changes over the culture period.

2 Materials

Prepare all solutions using ultrapure water and cell culture-grade reagents.

2.1 Maintenance of Human iPSC

1. Human iPSC line.
2. Phosphate-buffered saline (PBS; 1×).
3. StemFit AK02N (ECAK02N, Reprocell).
4. Y-27632 (034-24024, Wako): Dissolve 25 mg Y-27632 in 7.4 mL water. Aliquot Y-27632 solution and store in −20 °C.
5. Easy iMatrix-511 (892011, Nacalai).
6. Accutase (AT104, Nacalai).
7. Six-well plate (353046, BD Falcon).
8. Puromycin (P8833, Sigma): Dissolve 10 mg puromycin in 10 mL water. Aliquot and store at −20 °C.

2.2 Generation of iPSC Line Stably Expressing Tet-Inducible MyoD1

1. Human iPSC line.
2. Six-well plate.
3. Opti-MEM (Invitrogen).
4. Doxycycline-inducible MyoD1-expressing piggyBac vector (Tet-MyoD).
5. PiggyBac transposase vector PBase II.
6. NEPA 21 (Nepagene).

2.3 Differentiation of Human iPSC

1. Phosphate-buffered saline (PBS; 1×).
2. StemFit AK02N (ECAK02N, Reprocell).
3. Y-27632 (034-24024, Wako): Dissolve 25 mg Y-27632 in 7.4 mL water. Aliquot Y-27632 solution and store in −20 °C.

4. Matrigel growth factor reduced (MG) (356230, BD): Always hand MG on ice.
5. Accutase (AT104, Nacalai).
6. 10 cm dish.
7. Primate ES cell medium (RCHEMD001A, Reprocell) (PECM).
8. αMEM (21444-05, Nacalai).
9. Collagen type I solution (891500, Nippi) (*see* **Note 1**).
10. Insulin solution (093-06351 m Wako): 10 mg/mL.
11. Retinoic acid (R2625, Sigma): Prepare 10 mM retinoic acid in chloroform. Aliquot retinoic acid solution and store at −20 °C.
12. Thyroid hormone (T6397, Sigma): Prepare 1 mg/mL thyroid hormone solution in ethanol/1 M HCL (4:1). Aliquot thyroid hormone solution and store at −20 °C.
13. SB431542 (198-16543, Wako): Dissolve 25 mg SB431542 in 1.3 mL of DMSO. Aliquot SB431542 solution and store at −20 °C.
14. Doxycycline (19088-71, Nacalai) (Dox): Dissolve 10 mg doxycyline in 10 mL water. Aliquot doxycycline solution and store at −20 °C.
15. Horse serum (H1138, Sigma).
16. 2-Mercaptoethanol (21985023, Sigma): 55 mM.
17. Glucose solution (A24940–10, Invitrogen): 200 g/L.

2.4 Electric Field Stimulation System and Motion Imaging Assays

1. C-Pace EM (IonOptics) (*see* **Note 2**).
2. C-Dish (IonOptics) (*see* **Note 3**).
3. Inverted microscope.
4. SI8000 Cell Motion Imaging System (SONY) (*see* **Note 4**).

3 Methods

3.1 Maintenance of Human iPSCs

1. Add 1 mL of Easy iMatrix-511 per well of a six-well plate, and incubate the plate in a CO_2 incubator for at least 30 min.
2. Remove the plate from the incubator and aspirate the Easy iMatrix-511 solution.
3. Add 1.5 mL of StemFit AK02 medium containing 10 μM Y-27632 to each well.
4. Thaw a cryovial of hiPSCs and collect the cells in a 1.5 mL conical tube containing 2 mL of StemFit AK02 medium with 10 μM Y-27632.
5. Centrifuge the cells at 1000 rpm for 5 min.

6. Resuspend the cell pellet in 2 mL of StemFit AK02 medium with 10 μM Y-27632.
7. Count the cells.
8. Plate 1×10^4, 2×10^4, and 3×10^4 hiPSCs in separate wells.
9. Place the plate in a CO_2 incubator.
10. Perform a media change every other day with 1.5 mL of StemFit AK02 medium without Y-27632.

3.2 Generation of an iPSC Line Stably Expressing Tet-Inducible MyoD1

1. Add 1 mL of Easy iMatrix-511 per well of a six-well plate and incubate the plate in a CO_2 incubator for 30 min.
2. Remove the plate from the incubator and aspirate the Easy iMatrix-511 solution.
3. Add 1.5 mL of StemFit AK02 medium containing 10 μM Y-27632 to each well.
4. Wash the hiPSCs with 1 mL of PBS.
5. Dissociate the hiPSCs with 0.5 mL of Accutase for 10 min in a CO_2 incubator.
6. Neutralize Accutase by adding 1 mL of StemFit AK02 medium with 10 μM Y-27632.
7. Collect the cell suspension in a 15 mL conical tube.
8. Centrifuge the cells at 1000 rpm for 5 min.
9. Resuspend the cells in 2 mL of Opti-MEM medium containing 10 μM Y-27632.
10. Transfer 1.0×10^6 cells to a 1.5 mL microfuge tube.
11. Centrifuge the cells at 1000 rpm for 5 min.
12. Resuspend the cells in 1 mL of Opti-MEM medium containing 1 μg of the piggyBac transposase vector PBase II and 5 μg of the Tet-MyoD-expressing piggyBac vector.
13. Perform electroporation using the NEPA 21 (Nepagene) system at 125 V and 5 ms pulse width.
14. Plate 1×10^3, 3×10^3, 1×10^4, 3×10^4, and 1×10^5 cells per well of a six-well plate.
15. Start drug selection by adding StemFit AK02 medium containing 0.5 μg/mL puromycin.

3.3 Differentiation of Human iPSCs (Fig. 1) (Reproduced from [Uchimura et al. 2021] with Permission from Elsevier [9])

1. Add 6 mL of 1% Matrigel in advanced MEM (aMEM) per 10 cm dish and incubate the dish in a CO_2 incubator for 30 min.
2. Remove the dish from the incubator and aspirate the Matrigel solution.
3. Add 6 mL of StemFit AK02 medium containing 10 μM Y-27632 to the dish.

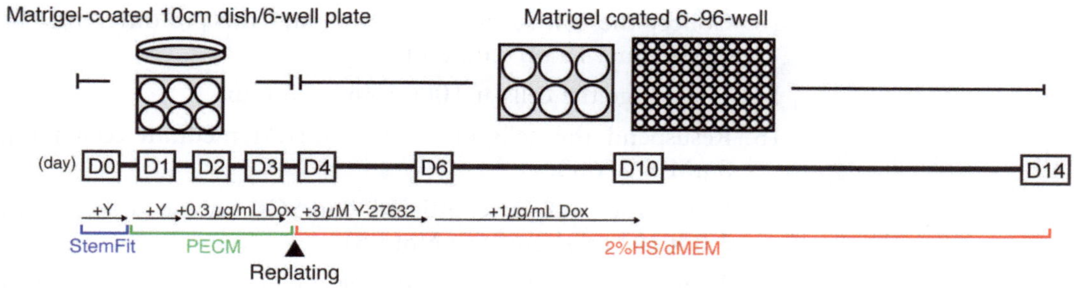

Fig. 1 Schematic of the hiPSC myogenic differentiation protocol. hiPSCs are pre-differentiated in primate ES cell medium (PECM) containing 0.3 mg/mL doxycycline. On day 4, pre-differentiated hiPSCs are replated in 2% horse serum medium with 3 μM Y-27632 without doxycycline. Doxycycline (1 mg/mL) is re-added to the 2% horse serum medium from day 6 for up to 4 additional days

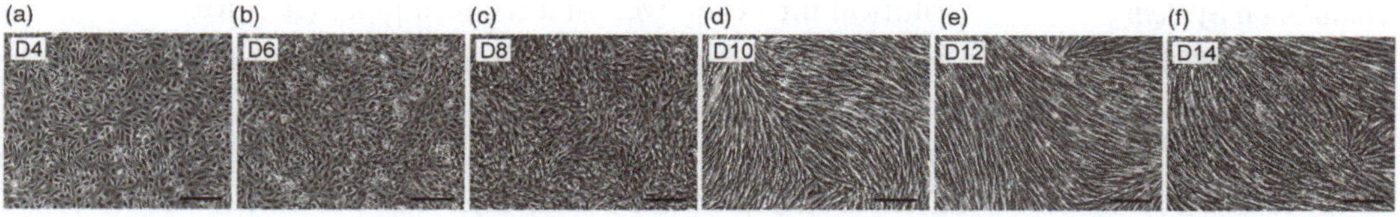

Fig. 2 Representative bright-field images showing the time-course of hiPSC myogenic differentiation from day 4 to day 14. Scale bar = 200 μm

4. Wash the hiPSCs with 1 mL of PBS.
5. Dissociate the hiPSCs with 0.5 mL of Accutase for 10 min in a CO_2 incubator.
6. Neutralize Accutase by adding 1 mL of StemFit AK02 medium with 10 μM Y-27632.
7. Collect the cell suspension in a 15 mL conical tube.
8. Centrifuge the cells at 1000 rpm for 5 min.
9. Plate the hiPSCs at a density of 3.0 × 104 cells/cm^2.
10. After 24 and 48 h, replace the media with PECM containing 10 μM Y-27632 and PECM containing 0.3 μg/mL doxycycline (Dox), respectively.
11. After an additional 48 h (corresponding to 4 days after the initial plating) (Fig. 2a) (Adapted from [Uchimura et al. 2021] with permission from Elsevier, [9]), dissociate the pre-differentiated hiPSCs with 3 mL of Accutase for 7 min in a CO2 incubator (referred to as replating).
12. Neutralize the cells with 3 mL of aMEM containing 2% horse serum, 200 μM 2-mercaptoethanol, 4.5 g/L glucose, 10 μg/mL insulin, 5 μM SB431542, 20 nM retinoic acid, 20 ng/mL thyroid hormones (2% HS/aMEM media), and 3 μM Y-27632 (*see* **Note 5**).

13. Collect the cell suspension containing the pre-differentiated hiPSCs in a 15 mL conical tube.
14. Centrifuge the cells at 1000 rpm for 5 min.
15. Resuspend the cells in 2% HS/aMEM medium containing 3 μM Y-27632.
16. Plate the pre-differentiated hiPSCs at a density of 5.0×10^4 cells/cm^2 (*see* **Note 6**).
17. After 48 h and 96 h, change the medium to 2% HS/aMEM supplemented with 1 μg/mL doxycycline (Fig. 2b–d) (Adapted from [Uchimura et al. 2021] with permission from Elsevier [9]).

3.4 Electric Field Stimulation System and Motion Imaging Assays (Fig. 3) (Reproduced from [Uchimura et al. 2021] with Permission from Elsevier [9])

1. Two days before replating, dispense 1.2 mL of collagen I solution into each well of a six-well plate (*see* **Note 7**).
2. Incubate the plate overnight in a CO_2 incubator to solidify the collagen gel.
3. One day before replating, dispense 2 mL of 1% Matrigel in aMEM on top of the collagen gel layer in each well.
4. Incubate the plate overnight in a CO_2 incubator.
5. Prepare pre-differentiated iPSCs as described previously (refer to **steps 1–8** in Subheading 3.3).
6. Aspirate the 1% Matrigel solution from the wells and add 2 mL of 2% HS/aMEM containing 10 μM Y-27632 before plating.
7. Plate the pre-differentiated hiPSCs at a density of 5.0×10^4 cells/cm^2 (*see* **Note 8**).
8. After 48 h, change the medium to 3 mL of 2% HS/aMEM containing 1 μg/mL doxycycline per well and transfer the plate to a 5% O_2 incubator (*see* **Note 9**).

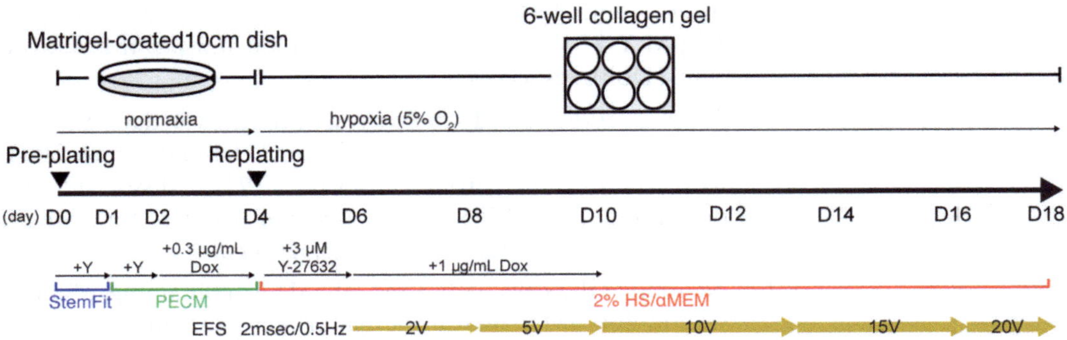

Fig. 3 Schematic of the hiPSC myogenic differentiation protocol with electric field stimulation (EFS). As in Fig. 1, hiPSCs are pre-differentiated and replated on day 4 onto collagen gel plates with replating conditions. Doxycycline (1 mg/mL) is re-added from day 6. EFS is initiated on day 6 at 2 V, 2 ms pulse, 0.5 Hz and increased to 5 V on day 8, 10 V on day 10, 15 V on day 13, and 20 V on day 16

9. Start stimulating the cells by electric field stimulation (EFS) using the C-Pace EM system and six-well C-Dish at 2 V with a 2 ms pulse interval and 0.5 Hz frequency continuously.
10. After 48 h, change the medium to fresh 3 mL of 2% HS/a-MEM containing 1 μg/mL doxycycline per well.
11. Increase the voltage to 5 V.
12. After 48 h, replace the medium with 3 mL of 2% HS/aMEM per well.
13. Increase the voltage to 10 V.
14. After 48 h, change the medium to fresh 3 mL of 2% HS/a-MEM per well.
15. After 24 h, increase the voltage to 15 V.
16. After 24 h, change the medium to fresh 3 mL of 2% HS/a-MEM per well.
17. After 48 h, change the medium to fresh 3 mL of 2% HS/a-MEM per well.
18. Increase the voltage to 20 V (*see* **Note 10**) (Fig. 4) (Reproduced from [Uchimura et al. 2021] with permission from Elsevier [9]).
19. Perform EFS and motion imaging assays as needed.
20. Record movies or time-lapse imaging using the SI8000 Motion Imaging System equipped with an Eclipse Ti inverted fluorescent microscope and a Stage Top Incubator (Tokai Hit) to maintain cells under humidified conditions at 37 °C and 5% CO_2 (*see* **Note 11**) (Fig. 5) (Reproduced from [Uchimura et al. 2021] with permission from Elsevier [9]).
21. The collagen gel plate was cultured in a Stage Top Incubator equipped with a C-Dish connected to the C-Pace EM (*see* **Note 12**).
22. Apply EFS at voltages between 10 and 20 V with a 2 ms pulse interval and 0.5–1 Hz frequency.

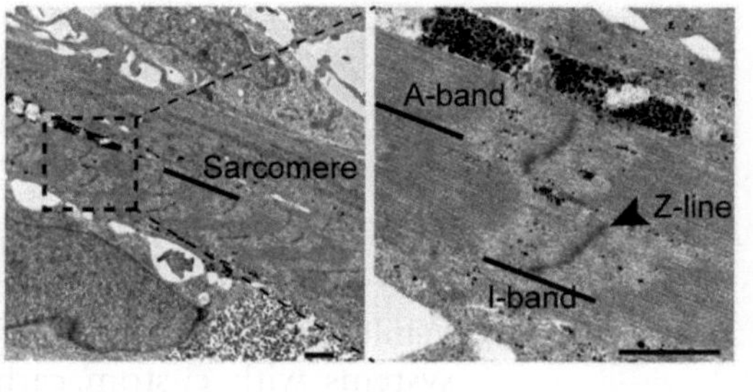

Fig. 4 Representative transmission electron microscopy images of myotubes differentiated and matured by 18 days of EFS. Rectangles indicate magnified areas. Scale bar = 3 μm

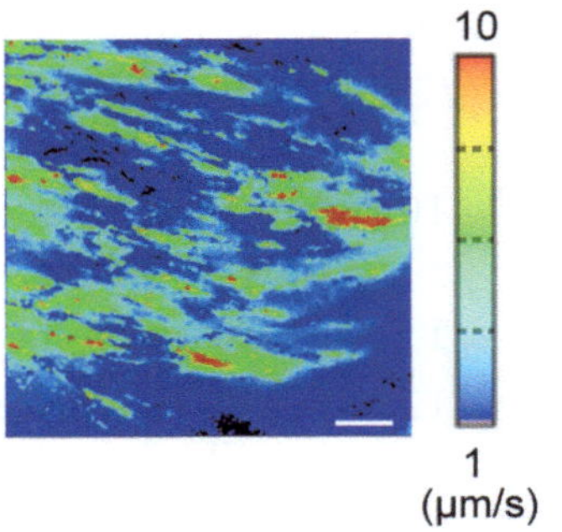

Fig. 5 Heatmap analysis of motion pixels in myotubes differentiated and matured by 18 days of EFS, acquired using the SI8000 Motion Imaging System. Scale bar = 100 μm

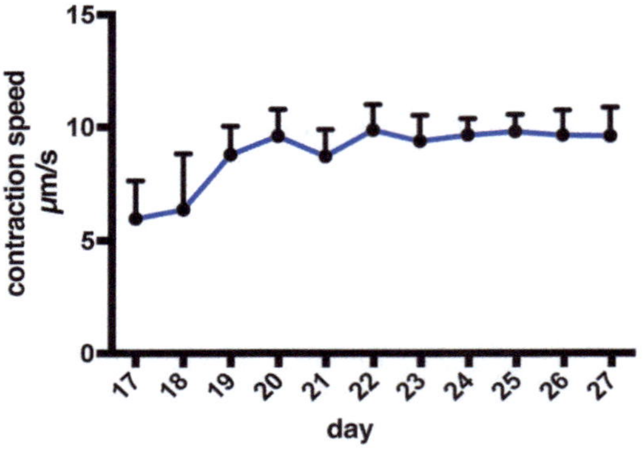

Fig. 6 Quantitative time-course analysis of myotube contraction from day 17 to day 27 under continuous EFS, analyzed using the SI8000 Motion Imaging System

23. Record movies or time-lapse imaging for 270 frames at a frame rate of 27 frames/second (equal to 10 s) and analyze using the associated software (*see* **Note 13**) (Fig. 6) (adapted from [Uchimura et al. 2021] with permission from Elsevier [9]).

4 Notes

1. The concentration of the collagen I solution for the gel should be optimized by researchers based on the specific skeletal muscle cell type being used.
2. Alternative electric field stimulation (EFS) systems can be used instead of the C-Pace EM system described.
3. While the C-Dish was used in this study, other cell stimulation systems with custom carbon or platinum electrodes can be employed.

4. The discontinued SI8000 Motion Imaging System can be replaced with confocal microscopy for time-lapse imaging. Imaging analysis can use software like ContractionWave [10] or MUSCLEMOTION [11] to quantify cell motion.
5. Titrate the Y-27632 dose (typical range 2–4 μM) to selectively allow survival of pre-differentiated but not undifferentiated hiPSCs.
6. Optimize the pre-differentiated hiPSC seeding density to achieve 100% confluency 48 h after replating.
7. Adjust collagen solution volume based on plate type, e.g., 1.0–1.5 mL per well of a six-well plate.
8. Adjust pre-differentiated hiPSC seeding density for the plate type to obtain confluency 2 days after replating.
9. Culture in 5% O_2 from this point onward to prevent excessive cell death during EFS culture.
10. Change the medium at least every other day, as EFS produces toxic byproducts. Adjust the volume and frequency based on cells and plate.
11. Record time-lapse imaging at 20–30 fps using an inverted confocal microscope.
12. The C-Dish and C-Pace can be substituted with custom carbon/platinum electrodes for EFS.
13. Use software like ContractionWave for motion analysis instead of the discontinued SI8000 software.

Conflict of Interest

The authors have no conflicts of interest to declare that are relevant to the content of this chapter.

References

1. Nikolic N, Gorgens SW, Thoresen GH, Aas V, Eckel J, Eckardt K (2017) Electrical pulse stimulation of cultured skeletal muscle cells as a model for in vitro exercise – possibilities and limitations. Acta Physiol (Oxf) 220(3): 310–331
2. Lee IH, Lee YJ, Seo H, Kim YS, Nam JO, Jeon BD et al (2018) Study of muscle contraction induced by electrical pulse stimulation and nitric oxide in C2C12 myotube cells. J Exerc Nutr Biochem 22(1):22–28
3. Park H, Bhalla R, Saigal R, Radisic M, Watson N, Langer R et al (2008) Effects of electrical stimulation in C2C12 muscle constructs. J Tissue Eng Regen Med 2(5): 279–287
4. Manabe Y, Miyatake S, Takagi M, Nakamura M, Okeda A, Nakano T et al (2012) Characterization of an acute muscle contraction model using cultured C2C12 myotubes. PLoS One 7(12):e52592
5. Nedachi T, Fujita H, Kanzaki M (2008) Contractile C2C12 myotube model for studying exercise-inducible responses in skeletal muscle. Am J Physiol Endocrinol Metab 295(5): E1191–E1204
6. Langelaan ML, Boonen KJ, Rosaria-Chak KY, van der Schaft DW, Post MJ, Baaijens FP

(2011) Advanced maturation by electrical stimulation: differences in response between C2C12 and primary muscle progenitor cells. J Tissue Eng Regen Med 5(7):529–539
7. Ito A, Yamamoto Y, Sato M, Ikeda K, Yamamoto M, Fujita H et al (2014) Induction of functional tissue-engineered skeletal muscle constructs by defined electrical stimulation. Sci Rep 4:4781
8. Raikova R, Krutki P, Celichowski J (2023) Skeletal muscle models composed of motor units: a review. J Electromyogr Kinesiol 70: 102774
9. Uchimura T, Asano T, Nakata T, Hotta A, Sakurai H (2021) A muscle fatigue-like contractile decline was recapitulated using skeletal myotubes from Duchenne muscular dystrophy patient-derived iPSCs. Cell Rep Med 2(6):100298
10. Scalzo S, Afonso MQL, da Fonseca NJ, ICG J, Alves AP, Mendonca C et al (2021) Dense optical flow software to quantify cellular contractility. Cell Rep Methods 1(4):100044
11. Sala L, van Meer BJ, Tertoolen LGJ, Bakkers J, Bellin M, Davis RP et al (2018) MUSCLEMOTION: a versatile open software tool to quantify cardiomyocyte and cardiac muscle contraction in vitro and in vivo. Circ Res 122(3):e5–e16

Chapter 14

Neurological Behavioral Assessment of Exon-Skipping Therapy for Neuromuscular Disorders Using Displaced Object Recognition Test, Open Field Test, and Three-Chamber Social Approach Test

Hiroya Ono and Yoshitsugu Aoki

Abstract

The displaced object recognition test (DORT), open field test (OFT), and three-chamber social approach test are standard behavioral assays used to assess the various aspects of cognition, emotional regulation, and social behavior in rodent models. This protocol introduces the use of these three tests to comprehensively evaluate the therapeutic effects of antisense oligonucleotide (ASO)-mediated exon-skipping therapies in neuromuscular disease models, such as Duchenne muscular dystrophy (DMD). Exon-skipping therapies aim to restore dystrophin expression, which can potentially influence cognitive, emotional, and social functions. By combining these tests, a multifaceted assessment of therapeutic efficacy can be achieved. This protocol is applicable not only to exon 51/53-skipping therapies but also to other exon-skipping treatments, enabling a broad evaluation of the therapeutic impact on cognition, emotional regulation, and social behavior.

Key words Displaced object recognition test (DORT), Open field test (OFT), Three-chamber social approach test, Exon-skipping therapy, Duchenne muscular dystrophy (DMD), Neuromuscular disorders, Cognitive impairment, Emotional regulation, Social behavior, Antisense oligonucleotides (ASO), Dystrophin isoform, Synaptic dysfunction, Therapeutic efficacy evaluation

1 Introduction

Duchenne muscular dystrophy (DMD) is an X-linked disorder caused by mutations in the DMD gene, leading to the absence of functional dystrophin protein and resulting in progressive muscle degeneration. Approximately 30% of DMD patients also present cognitive and behavioral impairments, with higher incidences of autism spectrum disorder (ASD) and attention-deficit/hyperactivity disorder (ADHD) [1, 2]. Additionally, emotional dysregulation and intellectual disability are common, and these impairments are strongly associated with the loss of dystrophin

Yoshitsugu Aoki and Norio Motohashi (eds.), *Splice Switching: Methods and Protocols*, Methods in Molecular Biology, vol. 2963, https://doi.org/10.1007/978-1-0716-4738-7_14,

isoforms, particularly Dp427 and Dp140, in the brain. Dp427 and Dp140 are known to influence synaptic integrity and neurotransmitter regulation, which are directly linked to the observed cognitive and behavioral deficits in DMD [3–5]. Notably, studies using the DMD model mouse (mdx52) suggest a relationship between Dp140 deficiency and ASD-like behaviors, providing critical insights into the behavioral abnormalities caused by dystrophin deficiency [6].

Antisense oligonucleotide (ASO)-mediated exon-skipping therapy is a promising approach for restoring functional dystrophin expression by correcting the reading frame of the DMD gene, allowing the production of truncated but functional dystrophin proteins [7]. While most studies have focused on muscle recovery, increasing attention is being given to the potential for exon-skipping therapy to alleviate cognitive, emotional, and social impairments caused by dystrophin deficiency in the brain.

In this protocol, we propose the use of the displaced object recognition test (DORT), open field test (OFT), and three-chamber social approach test to assess the therapeutic effects of exon-skipping therapy in DMD models. DORT evaluates spatial memory and cognition, OFT measures spontaneous activity and anxiety-related behaviors, and the three-chamber social approach test assesses sociability and social novelty. By integrating these three tests, the therapeutic impact of exon-skipping therapy on cognitive, emotional, and social functions can be comprehensively evaluated.

2 Displaced Object Recognition Test (DORT) (Fig. 1)

2.1 Materials

1. Apparatus: A large, opaque plastic storage bin (dimensions approximately 40 cm × 40 cm × 30 cm) with four small bottle caps securely attached to the bottom of the bin to hold objects in place. The opaque walls prevent the subject from being distracted by external stimuli, ensuring focused interaction with the objects.
2. Objects: Two distinct objects (wide-mouth opaque amber plastic bottles, 250 mL) are chosen so that they are easily distinguishable by both visual and tactile cues to the subject. It is important that the objects do not have a scent that could influence the mouse's behavior.
3. Bedding: Corn cob bedding spread evenly across the bottom of the apparatus to mask olfactory cues and create a uniform surface for the subject's exploration. Corn cob bedding is specifically chosen to provide a neutral olfactory and tactile environment, minimizing external distractions during object exploration. It also provides a comfortable substrate that prevents stress-related behaviors.

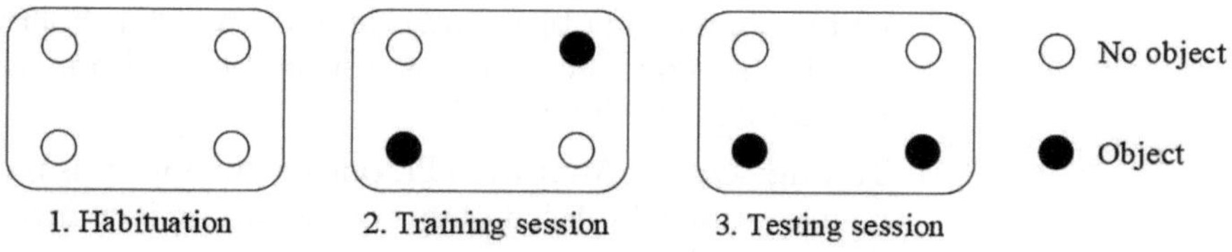

Fig. 1 Displaced object recognition test (DORT)
1. The habituation session allows mice to acclimate to the testing apparatus. Mice are placed in an empty arena without objects for 10 min per day over two consecutive days. This step reduces stress and ensures familiarity with the environment before the introduction of objects during subsequent sessions
2. The training session involves placing two identical objects in the testing apparatus. Mice explore objects freely for 10 min. Exploration is defined as the subject's nose being within 2 cm of an object and oriented toward it. This session establishes a baseline for object interaction and familiarity
3. In the testing session, one object is moved to a new location, creating a "displaced object." Mice are reintroduced to the apparatus and allowed to explore for 10 min. The time spent exploring the displaced versus stationary objects is recorded to assess spatial memory, with a preference for the displaced object indicating successful recognition

4. Tracking Software: An automated tracking system, such as EthoVision (Noldus, Netherlands) or ANY-Maze (Stoelting Co., USA), should be used to precisely measure the subject's exploration time and movement patterns around the objects. It allows for real-time, high-resolution tracking of the mouse's behavior, minimizing human bias.
5. Environmental Control: The temperature of the testing room should be maintained between 21 and 24 °C, and the relative humidity should be controlled at 40–60%. Consistent lighting is essential; ideally, overhead fluorescent lighting should be used to minimize shadows, which could influence the mouse's behavior. The light intensity should be kept low to medium (200–400 lux) to avoid undue stress.
6. Cleaning Supplies: Use 70% ethanol to clean the apparatus and objects between trials to eliminate olfactory cues that might influence subsequent trials. Thorough cleaning is crucial for ensuring that mice are not using scent-based strategies to explore the objects.

2.2 Methods [8]

1. Habituation: The subject mice are first habituated to the testing apparatus without objects for 10 min per day over two consecutive days. This allows the animals to become familiar with the environment and reduces stress-related behavior during testing.
2. Training Session: On the third day, two identical objects are placed in the apparatus, and the mouse is allowed to explore for 10 min. Exploration is defined as the subject's nose being

within 2 cm of an object and oriented toward it. Following the training session, the mouse is returned to its home cage for a 5-min inter-trial interval (ITI).

3. Testing Session: After the ITI, one of the objects is moved to a new location (displaced object). The mouse is reintroduced to the apparatus and given 10 min to explore. The time spent exploring the displaced and stationary objects is recorded. A preference for the displaced object indicates successful spatial memory.
4. Data Analysis: Exploration time spent with each object is measured using an automated tracking system. A discrimination index is calculated as (time spent with displaced object - time spent with stationary object) / total time spent with both objects, providing a quantitative measure of memory.

3 Open Field Test (OFT)

3.1 Materials

1. Apparatus: A large square or circular open field arena (typically 40 cm × 40 cm for a square arena or 40 cm in diameter for a circular one), with high walls (at least 30 cm) to prevent the subject from escaping. The floor of the arena can be marked with a grid to help quantify movement. If a square arena is used, the center is typically defined as a smaller square (e.g., 10 cm × 10 cm) in the middle.
2. Tracking Software: An automated tracking system should be used to monitor the mouse's movements. The software will record metrics like the total distance traveled, time spent in different zones (e.g., the center versus the periphery), and specific behaviors such as freezing, rearing, and grooming. The use of automated systems ensures high reproducibility and precision, reducing the risk of observer bias.
3. Environmental Control: Testing should be conducted in a room kept at a temperature between 21 and 24 °C, with humidity maintained at 40–60%. Lighting should be bright (400–600 lux) but even, to introduce mild stress that encourages the display of anxiety-like behaviors. Avoiding shadows or changes in lighting is crucial to prevent the subject from being distracted or overly stressed by the environment.
4. Cleaning Supplies: Nonscented cleaning agents should be used to clean the open field arena thoroughly between trials to remove olfactory traces from previous subjects, which could otherwise influence behavior. Cleaning should focus on all surfaces with which the subject may come into contact.

3.2 Methods

1. Habituation: Each subject mouse is introduced to the testing room 30 min before the start of the experiment for acclimatization.
2. Testing Session: The mouse is placed in the center of the open field arena and allowed to explore freely for 10 min. Spontaneous locomotor activity and anxiety-related behavior, such as time spent in the center versus the periphery, are measured. Parameters recorded include the total distance traveled, the time spent in the center area, and the number of rearing events. The definitions for observed behaviors are as follows [9, 10]:
   - Exploration:

     Behavior involves the mouse sniffing the floor or walls. During the exploratory behavior, the mouse exhibits relaxed and natural movements.

     Criteria: The mouse's nose is in contact with or within 1–2 cm of the object or surface, and its body is oriented toward the target.
   - Time in Center Area:

     The duration the mouse spends in the central area (e.g., 10 cm × 10 cm square) of the arena. This metric is used to assess the anxiety-related behavior.

     Criteria: The mouse's head and torso (body axis) must be entirely within the center area. The tail alone entering the center area does not count.
   - Time in Periphery Area:

     The duration the mouse spends moving along or staying close to the walls of the arena. Increased time in the periphery typically indicates heightened anxiety.

     Criteria: The mouse's head and torso are within the periphery area, often observed as wall-following behavior.
   - Rearing:

     Behavior where the mouse stands on its hind legs, extending its forelimbs either against the wall or into the air. This is often an exploratory behavior.

     Criteria: The mouse's forelimbs are raised off the floor, with only the hind limbs remaining in contact with the ground.
   - Freezing:

     A complete cessation of movement where the mouse remains immobile, often indicating fear or heightened anxiety.

     Criteria: The mouse remains entirely still (excluding subtle movements like breathing or slight ear twitches) for at least 2 s.

- Grooming:

  Behavior where the mouse uses its forelimbs or mouth to clean its face, body, or tail. Grooming is often associated with stress relief or adaptation to the environment.

  Criteria: The mouse is observed licking or rubbing its body parts with its forelimbs or mouth.

- Total Distance Traveled:

  The cumulative distance the mouse moves within the arena during the test session.

  Criteria: This parameter is recorded using automated tracking software based on the position of the mouse's center of gravity.

3. Environmental Controls: The room temperature is kept between 21 and 24 °C, and the lighting is set to 400–600 lux to introduce mild stress, as rodents generally prefer darker environments. This stress is intentional to elicit anxiety-like behavior and enhance the sensitivity of the test.
4. Data Analysis: The automated tracking system records movement parameters, and anxiety-related behavior is assessed by the percentage of time spent in the center versus the periphery. Increased time in the periphery is interpreted as heightened anxiety.

4 Three-Chamber Social Approach Test (Fig. 2)

4.1 Materials

1. Apparatus: A three-chamber setup, typically made from transparent materials like acrylic to allow for visual monitoring of the subject. Each chamber should be approximately 20 cm × 40 cm, and small doors (8 cm wide) between the chambers should allow the subject to move freely between them. The doors should have no obstructions, allowing easy movement for the mouse.
2. Wire Cages: Two small wire cages (typically cylindrical, with dimensions around 10 cm in diameter and 15 cm in height) are placed in the side chambers. These cages are designed to contain target mice for visual, auditory, and olfactory interaction without allowing direct physical contact. The wire spacing should be narrow enough to prevent biting but wide enough to allow olfactory cues to pass.
3. Subject and Target Mice: The subject mice should be exon-skipping therapy models (e.g., mdx or mdx52), aged between 6 and 10 weeks. Target mice should be age- and sex-matched but unfamiliar to the subject mouse, as this encourages normal social interaction behaviors. It is important to standardize the conditions of both subject and target mice to avoid confounding variables.

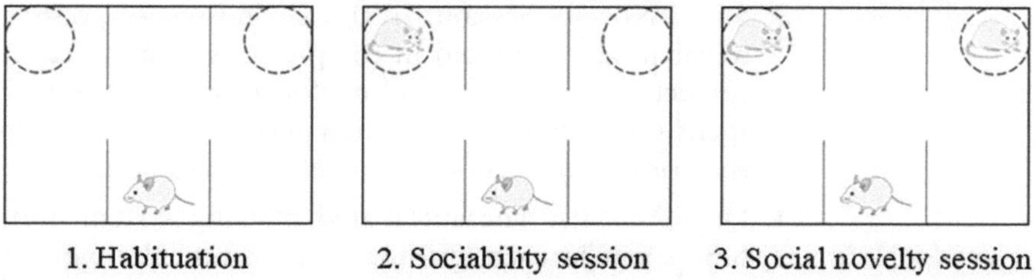

Fig. 2 Three-chamber social approach test
1. Habituation: On the day prior to testing, each mouse is habituated to the apparatus for 10 min, with unrestricted access to all chambers
2. Sociability session: On the testing day, the mouse is placed in the center chamber with access to both side chambers. One side chamber contains a novel mouse enclosed in a wire cage, while the other contains an empty wire cage. The mouse is allowed to explore all three chambers for 10 min, and the time spent near the novel mouse is measured as an indicator of sociability
3. Social novelty session: In a second 10-min session, a second unfamiliar mouse is placed in the previously empty wire cage. The time spent interacting with the novel mouse versus the familiar mouse is recorded to assess social novelty recognition

4. Tracking Software: Automated tracking system is recommended for recording movement across the chambers. It tracks the time the subject mouse spends in each chamber, as well as the frequency and duration of its interactions with the target mice, providing detailed and objective data.
5. Environmental Control: The testing room should be kept at 21–24 °C with humidity levels maintained at 40–60%. Lighting should be bright but consistent across all chambers to avoid distracting shadows or hot spots. The arena should be free of any outside noise to minimize stress or distractions that could alter behavior.

4.2 Methods [11, 12]

1. Habituation: Each subject mouse is acclimated to the apparatus for 10 min on the day prior to testing, with all doors between chambers open.
2. Sociability Session: On the testing day, the mouse is placed in the center chamber, with the doors to both side chambers open. One side chamber contains a novel mouse (enclosed in a wire cage), while the other chamber contains an empty wire cage. The subject mouse is allowed to explore all three chambers for 10 min. The amount of time spent in proximity to the novel mouse is recorded, indicating sociability. During the sociability session, additional behaviors such as sniffing duration, proximity maintenance, and chamber transition frequency should be analyzed to provide a more comprehensive evaluation of social interaction dynamics.

3. Social Novelty Session: In the second session, a second unfamiliar mouse is placed in the previously empty cage, and the subject is allowed to explore for another 10 min. The time spent interacting with the new mouse versus the familiar one is recorded to assess social novelty recognition.
4. Data Analysis: The automated tracking system measures the time spent in each chamber and around the wire cages. The preference for social interaction and recognition of social novelty is evaluated by comparing the time spent in proximity to the novel versus the familiar mouse.

This method provides a robust and unbiased assessment of cognitive, emotional, and social impairments in the mouse models of neuromuscular disorders. The integration of automated tracking systems ensures objectivity and high reproducibility in the data collection process.

5 Notes

1. Handling and Stress [13]: Prior to all tests, gentle handling of the mice is essential. Mice should be handled for several days before testing to reduce stress-induced behaviors during the tests, especially in anxiety-related assessments like the open field test (OFT). Avoid handling the subjects on the day of testing to prevent heightened stress levels.
2. Environmental Consistency [14]: Ensure that all environmental conditions, such as temperature, humidity, and lighting, remain consistent across all trials. Even small fluctuations in temperature or lighting can affect the mouse behavior, introducing variability in the data. Testing should also be conducted at the same time of day for all subjects to avoid circadian influences on behavior.
3. Automated Tracking Behavior Systems [15]: Automated tracking systems, such as EthoVision (Noldus, the Netherlands) or ANY-Maze (Stoelting Co., USA), ensure precision and consistency across trials, significantly reducing inter-observer variability. These systems are highly recommended for their ability to provide comprehensive behavioral metrics.
4. Session Duration [16]: To prevent fatigue or disinterest, it is essential to limit each test session to 10–15 min. Longer sessions can lead to reduced exploratory behavior, increased stress, or altered social interactions, which could confound the data.
5. Apparatus Cleaning [17]: After each session, all apparatuses and objects should be thoroughly cleaned with 70% ethanol to eliminate olfactory cues that could affect the behavior of

subsequent subjects. Objects used in social and recognition tests should be rotated between sessions to minimize any potential bias from lingering scent marks or visual wear.

Conflict of Interest

The authors have no conflicts of interest to declare that are relevant to the content of this chapter.

References

1. Bresolin N et al (1994) Cognitive impairment in Duchenne muscular dystrophy. Neuromuscul Disord 4(4):359–369
2. Wicksell RK et al (2004) Specific cognitive deficits are common in children with Duchenne muscular dystrophy. Dev Med Child Neurol 46(3):154–159
3. Bardoni A et al (2000) Loss of Dp140 regulatory sequences is associated with cognitive impairment in dystrophinopathies. Neuromuscul Disord 10(3):194–199
4. Felisari G et al (2000) Loss of Dp140 dystrophin isoform and intellectual impairment in Duchenne dystrophy. Neurology 55(4): 559–564
5. Chamova T et al (2013) Association between loss of Dp140 and cognitive impairment in Duchenne and Becker dystrophies. Balkan J Med Genet 16(1):21–30
6. Hashimoto Y et al (2022) Brain Dp140 alters glutamatergic transmission and social behaviour in the mdx52 mouse model of Duchenne muscular dystrophy. Prog Neurobiol 216: 102288
7. Yokota T, Duddy W, Partridge T (2007) Optimizing exon skipping therapies for DMD. Acta Myol 26(3):179–184
8. McLagan AN, Hales JB (2019) Displaced object recognition memory in rats. Bio Protoc 9(8):e3212
9. Crawley JN(2007) Emotional behaviors: animal models of psychiatric diseases. In: What's wrong with my mouse? pp 226–265
10. Seibenhener ML, Wooten MC (2015) Use of the open field maze to measure locomotor and anxiety-like behavior in mice. J Vis Exp 96: e52434
11. Nadler JJ et al (2004) Automated apparatus for quantitation of social approach behaviors in mice. Genes Brain Behav 3(5):303–314
12. Kaidanovich-Beilin O et al (2011) Assessment of social interaction behaviors. J Vis Exp 48: 2473
13. Gouveia K et al (2017) Optimising reliability of mouse performance in behavioural testing: the major role of non-aversive handling. Sci Rep 7: 44999
14. Saré RM et al (2021) Behavior Testing in Rodents: Highlighting Potential Confounds Affecting Variability and Reproducibility. Brain Sci 11(4): 522
15. Lim CJM et al (2023) Comparison of automated video tracking systems in the open field test: ANY-Maze versus EthoVision XT. J Neurosci Methods 397:109940
16. Sestakova N et al (2013) Determination of motor activity and anxiety-related behaviour in rodents: methodological aspects and role of nitric oxide. Interdiscip Toxicol 6(3):126–35
17. Hershey JD et al (2018) Effects of Various Cleaning Agents on the Performance of Mice in Behavioral Assays of Anxiety. J Am Assoc Lab Anim Sci 57(4): 335-339

Chapter 15

Analysis of Prefrontal–Amygdala Synaptic Functions Using Optogenetics and Patch-Clamp Recording

Hiroshi Kuniishi and Eri Takeuchi

Abstract

Duchenne muscular dystrophy (DMD) is an X-linked neuromuscular disease characterized by progressive muscle wasting caused by a mutation in the *DMD* gene encoding the dystrophin protein. Dystrophin is highly expressed in the central nervous system (CNS), and DMD patients are often diagnosed with abnormalities in emotional and social functions. However, the pathophysiology of the CNS in DMD patients is not fully understood. The DMD mouse models show abnormalities in anxiety, fear and social behavior, and synaptic dysfunction in the amygdala, which is involved in emotional and social processing. The direct input from the prefrontal cortex (PFC) to the amygdala (PFC–amygdala pathway) plays a crucial role in emotional and social processing. Hence, examination of the PFC–amygdala synaptic function is important for not only investigating the pathophysiology of related psychiatric symptoms but also for assessing the efficacy of novel treatments. Here, we present a protocol of an electrophysiological method combined with optogenetics to examine the PFC–amygdala synaptic functions in mice.

Key words Electrophysiology, Optogenetics, Amygdala, Prefrontal cortex

1 Introduction

Duchenne muscular dystrophy (DMD) is an X-linked neuromuscular disease characterized by progressive muscle wasting caused by a mutation in the *DMD* gene which encodes dystrophin protein [1]. In addition to muscular symptoms, DMD patients are often present with cognitive deficits and psychiatric symptoms including learning disabilities, elevated anxiety, and autism spectrum disorder (ASD) [2–5]. Hence, understanding pathophysiology in the central nervous system, in addition to neuromuscular dysfunctions, is a significant challenge for the development of treatments for DMD.

Interestingly, dystrophin is highly expressed in the CNS, especially in the amygdala [6]. The amygdala is one of the key structures for the regulation of anxiety, fear, and social behavior, and its dysfunction is associated with emotional and social deficits [7, 8]. Previously, we have demonstrated that DMD mouse models

Yoshitsugu Aoki and Norio Motohashi (eds.), *Splice Switching: Methods and Protocols*, Methods in Molecular Biology, vol. 2963,
https://doi.org/10.1007/978-1-0716-4738-7_15,

show abnormalities in anxiety, fear and social behavior, and amygdala synaptic transmission. Full-length dystrophin (Dp427) and shorter isoforms, Dp140 and Dp71, are expressed in the brain. In Dp427-deficient dystrophic *mdx* mice, the number of $GABA_A$ receptor clusters in the basolateral amygdala (BLA) was reduced, and GABAergic transmission of the BLA was impaired [9]. The *mdx* mice also showed enhanced fear response [9] and anxiety associated with acute mild stress [10]. The *mdx*52 mouse model, deficient in both Dp427 and Dp140, also exhibits these BLA-related behavioral and synaptic phenotypes [11].

Anatomically, the amygdala receives dense input from the prefrontal cortex (PFC), and recent studies have highlighted that the PFC is thought to play a critical role in emotion and social processing through top-down regulation of the amygdala [12–14]. Previous studies demonstrated that artificial manipulation of PFC–amygdala transmission alters fear-related and social behaviors [15–18]. In addition, changes in PFC–amygdala synaptic properties were observed in mice with fear conditioning or social deficits [19–21]. Furthermore, our previous study demonstrates that *mdx*52 mice exhibit abnormal PFC–BLA excitatory synaptic transmission and social behavior, and these abnormalities were rescued by exon-skipping or mRNA-based drugs [11].

Together, the examination of PFC–amygdala synaptic function is important for not only investigating the pathophysiology of psychiatric symptoms with emotional and social deficits but also assessing the efficacy of novel treatments. Here, we present our protocol of the electrophysiological methods using optogenetics to examine the synaptic function in the PFC–BLA pathway.

2 Materials

A combination of adeno-associated virus (AAV)-driven channelrhodopsin2 (ChR2) expression and patch-clamp recording allows isolation of input-specific synaptic responses. This method is useful to examine synaptic properties in long-range projection and thus is ideal for the PFC–amygdala circuit. Recently, we revealed abnormalities in excitatory synapse in pathways projecting from the prefrontal subregion (medial PFC [mPFC] or medial orbitofrontal cortex [mOFC]) to BLA in *mdx*52 mice or adolescent socially isolated mice that display abnormal social behavior by using this method [11, 16]. The detailed materials and methods and procedures are described in the following subheadings.

2.1 AAV Microinjection into Mice Brains

1. Stereotaxic instrument and manipulator (SR-6 M-HT, SM-15, Narishige, Tokyo, Japan) and micro infusion pump (UMP-3, World Precision Instruments, FL, USA).
2. Ketamine/xylazine cocktails for anesthesia.

3. AAV solution (AAV5-CaMKIIa-ChR2-EYFP, obtained from vector core at the university of North Carolina at Chapel Hill) and 10 μL Hamilton microsyringe.

2.2 Preparation of Acute Brain Slices and Electrophysiological Recording

1. Artificial cerebrospinal fluid (aCSF; containing 125 mM NaCl, 4.4 mM KCl, 1.5 mM $MgSO_4$, 1.0 mM NaH_2PO_4, 26 mM $NaHCO_3$, 10 mM glucose, 2.5 mM $CaCl_2$; pH 7.4, 290–300 mOsm/L).
2. Sevoflurane for anesthesia.
3. Linear slicer for acute brain slice (Dosaka EM Co., Ltd., Kyoto, Japan).
4. Electrophysiology rig for patch-clamp recording (microscope equipped with IR-DIC [DM LFSA, Leica, Germany], CCD-camera and monitoring system [IR-1000, DAGE-MIT, IN, USA], patch-clamp amplifier [MultiClamp 700B, Axon Instruments, Union City, CA, USA] and data digitizer [Digidata 1550B, Axon Instruments], micromanipulator [PatchStar, Scientifica, UK], vibration isolation table, and perfusion system with in-line heater).
5. High-power LED light source (LEX-2B, BrainVision, Tokyo, Japan) for activation of ChR2 and excitation of EYFP. Stimulator (SEN-7203, Nihon Kohden, Tokyo, Japan) for TTL trigger for control light stimulation.
6. Borosilicate glass capillary and pipette puller (PC-10, Narishige).
7. Cs-based patch electrode solution containing 105 mM CsOH, 30 mM CsCl, 10 mM HEPES, 0.5 mM EGTA, 1 mM $MgCl_2$, 12 mM Na-phosphocreatine, 3 mM Mg-ATP, 0.5 mM Na-GTP, 0.1 mM spermine, 1 mM QX-314, pH 7.3 with gluconic acid, 295 mOsm/L.
8. Data acquisition software package (ClampX software, Axon Instruments) for data analysis.

3 Methods

3.1 Preparation of Animals

1. Fill the microsyringe with AAV solution and set the syringe to infusion pump and stereotaxic manipulator.
2. Anesthetize the mice (3–5 weeks of age) by intraperitoneal injection of ketamine (100 mg/kg) and xylazine (20 mg/kg) and fix on a stereotaxic instrument.
3. Remove a small segment of the skull above mPFC (AP +1.98 mm, ML ±0.2 mm from bregma) or mOFC (AP +2.7 mm, ML ±0.25 mm from bregma) using a micro drill.

4. Insert the microsyringe needle into the depth of the targeted brain region (mPFC: DV −1.2 mm from dura, mOFC: DV −1.5 mm from dura) and slowly inject AAV solution under the control of an infusion pump (*see* **Note 1**). After viral injection, slowly retract the needle from the brain, and suture closed the scalp.
5. Four weeks after injection, perform electrophysiological experiments using the viral-injected mice (Fig. 1a).

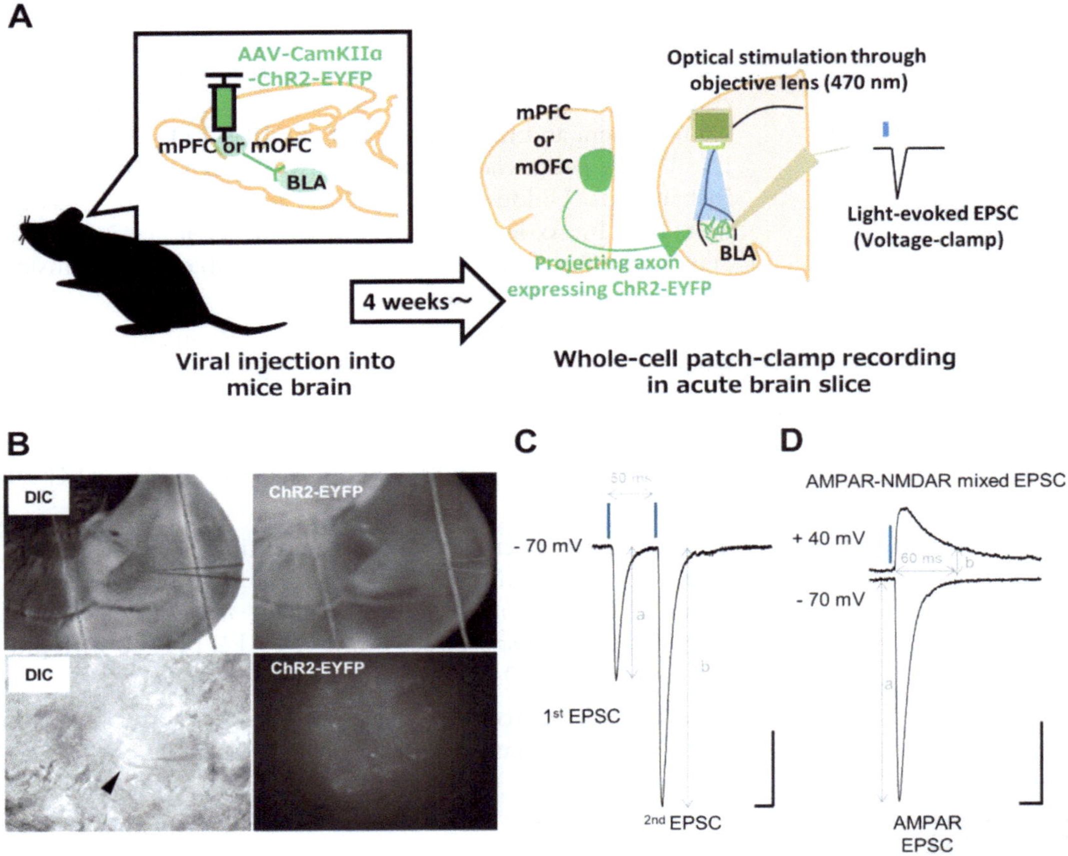

Fig. 1 (**a**) Schematic representations of experimental procedure of optogenetic isolation of prefrontal cortex (PFC)–basolateral amygdala (BLA) synaptic transmission. (**b**) Representative images of recorded neurons in BLA magnified by 4x (top) and 63x objective lens (bottom). Cell contour is visualized with DIC (left), and ChR2-EYFP expressing axons is visualized with GFP filters (right). The arrowhead indicates the soma of the recorded neuron. (**c**) Representative trace of LE-EPSC evoked by paired set of blue light stimulation in PFC–BLA synapse (50 ms interval). Blue lines indicate light stimulation. Scale: 100 pA, 20 ms. The paired-pulse ratio is calculated as the ratio of amplitude of second (**b**) and first LE-EPSCs (**a**). (**d**) Representative trace of LE-EPSC evoked by single blue light application in the PFC–BLA synapse at −70 mV (bottom) and + 40 mV holding potentials (top). Blue lines indicate light stimulation. Scale: 100 pA, 20 ms. The AMPA/NMDA current ratio is calculated as the ratio of amplitude of AMPAR-mediated (**a**) and NMDAR-mediated LE-EPSCs (**b**)

3.2 Preparation of Acute Brain Slices

1. For brain slicing, prepare aCSF and keep ice-cold before slice preparation.
2. For brain slice incubation, prepare fresh aCSF at the recording day, and keep room temperature.
3. Anesthetize the mice with sevoflurane and decapitate. Quickly remove the brain and put it in the ice-cold aCSF bubbling with 5% CO_2 and 95% O_2 mixed gas.
4. Slice the brain to 300 μm thickness section containing amygdala in ice-cold aCSF using a linear slicer (Dosaka EM Co., Ltd., Kyoto, Japan). For recovery, transfer the slices to fresh aCSF bubbling with the mixed gas and incubate at room temperature for at least 30 min.
5. Slice the brain region containing AAV-injected site (mPFC or mOFC) and confirm the expression of ChR2-EYFP using fluorescence microscopy.

3.3 Whole-Cell Patch Clamp Recordings in Brain Slices and Optogenetic Isolation of the PFC–BLA Synaptic Response

1. After recovery, transfer the brain slice to the recording chamber perfused (3.0 mL/min, gravity flow) with fresh aCSF maintained at 28–32 °C with an in-line heater and an automatic temperature controller.
2. Confirm the ChR2-EYFP expression in prefrontal projecting axon fiber in the BLA with blue LED light source and GFP or EYFP filter (Fig. 1b).
3. Whole-cell patch-clamp recordings were taken from BLA principal neurons with pyramidal morphology surrounded by ChR2-expressing axon (Fig. 1b). The electrophysiological signal is amplified and filtered at 5 kHz using a MultiClamp 700B patch-clamp amplifier. Data are digitized at 50 kHz and acquired using Clampex software.
4. Under the voltage clamp at −70 mV, apply blue LED light to the microscopic field around the recorded cell. In response to stimulation, light-evoked EPSC (LE-EPSC) is recorded (*see* **Note 2**).
5. To assess excitatory presynaptic short-term plasticity, measure the paired-pulse ratio(Fig. 1c). Under perfusion with $GABA_AR$ antagonist, picrotoxin (100 uM in aCSF, *see* **Note 3**), apply paired pulse of blue light stimulation at various intervals (e.g., 50 ms, 100 ms, 250 ms, 500 ms, and 1000 ms). The paired-pulse ratio is calculated as a ratio of the second to the first LE-EPSCs (*see* **Note 4**).
6. To assess the postsynaptic changes in synaptic strength, measure the AMPA/NMDA current ratio (Fig. 1d) [22, 23]. Under perfusion with picrotoxin, record AMPA receptor-mediated inward LE-EPSC at −70 mV holding potential. Then, record the AMPA and NMDA receptor-

mixed outward LE-EPSC at +40 mV holding potential (*see* **Note 5**). Obtain an amplitude of 60 ms following light stimulation for NMDA-mediated LE-EPSC (*see* **Note 6**). As the AMPA/NMDA current ratio, calculate the ratio of AMPA receptor-mediated LE-EPSC to NMDA receptor-mediated LE-EPSC.

4 Notes

1. In our experiments, infusion volume and rate are set at 300 nL and 100 nL/min, respectively. After the injection, leave the injection needle at the site for 10 min.
2. Intensity of the light stimulation is 1.8–70.7 × 10–7 J under the objective lens in our equipment [24].
3. Apply picrotoxin for blockade-inhibitory synaptic transmission and isolate excitatory synaptic transmission.
4. Generally, the amplitude of second LE-EPSC is greater than first LE-EPSC (paired-pulse facilitation) in mPFC-BLA or mOFC-BLA synapses at 50 ms to 250 ms interval [11, 16].
5. Cs-based internal pipette solution allows to record outward EPSC at +40 mV holding potential.
6. NMDAR-mediated current is isolated because AMPA current declines within 60 ms from the light onset.

Acknowledgments

This research was supported by JSPS KAKENHI (grant number: 18 K15533 and 21 J01636) and research grant from Kawano Masanori Memorial Public Interest Incorporated Foundation for Promotion of Pediatrics, Japan. We would like to thank Dr. Chaitra Sathyaprakash for helpful comments and English editing.

Conflict of Interest The authors have no conflicts of interest to declare that are relevant to the content of this chapter.

References

1. Hoffman EP, Brown RH Jr, Kunkel LM (1987) Dystrophin: the protein product of the Duchenne muscular dystrophy locus. Cell 51(6):919–928
2. Cotton S, Voudouris NJ, Greenwood KM (2001) Intelligence and Duchenne muscular dystrophy: full-scale, verbal, and performance intelligence quotients. Dev Med Child Neurol 43(7):497–501. https://doi.org/10.1017/s0012162201000913
3. Banihani R, Smile S, Yoon G, Dupuis A, Mosleh M, Snider A et al (2015) Cognitive and neurobehavioral profile in boys with Duchenne muscular dystrophy. J Child Neurol

30(11):1472–1482. https://doi.org/10.1177/0883073815570154

4. Ricotti V, Mandy WP, Scoto M, Pane M, Deconinck N, Messina S et al (2016) Neurodevelopmental, emotional, and behavioural problems in Duchenne muscular dystrophy in relation to underlying dystrophin gene mutations. Dev Med Child Neurol 58(1):77–84. https://doi.org/10.1111/dmcn.12922
5. Saito Y, Takeshita E, Komaki H, Nishino I, Sasaki M (2022) Determining neurodevelopmental manifestations in Duchenne muscular dystrophy using a battery of brief tests. J Neurol Sci 440:120340. https://doi.org/10.1016/j.jns.2022.120340
6. Doorenweerd N, Mahfouz A, van Putten M, Kaliyaperumal R, PAC TH, JGM H et al (2017) Timing and localization of human dystrophin isoform expression provide insights into the cognitive phenotype of Duchenne muscular dystrophy. Sci Rep 7(1):12575. https://doi.org/10.1038/s41598-017-12981-5
7. Adolphs R (2010) What does the amygdala contribute to social cognition? Ann N Y Acad Sci 1191(1):42–61. https://doi.org/10.1111/j.1749-6632.2010.05445.x
8. Janak PH, Tye KM (2015) From circuits to behaviour in the amygdala. Nature 517(7534):284–292. https://doi.org/10.1038/nature14188
9. Sekiguchi M, Zushida K, Yoshida M, Maekawa M, Kamichi S, Yoshida M et al (2009) A deficit of brain dystrophin impairs specific amygdala GABAergic transmission and enhances defensive behaviour in mice. Brain 132(Pt 1):124–135. https://doi.org/10.1093/brain/awn253
10. Vaillend C, Chaussenot R (2017) Relationships linking emotional, motor, cognitive and GABAergic dysfunctions in dystrophin-deficient mdx mice. Hum Mol Genet 26(6):1041–1055. https://doi.org/10.1093/hmg/ddx013
11. Hashimoto Y, Kuniishi H, Sakai K, Fukushima Y, Du X, Yamashiro K et al (2022) Brain Dp140 alters glutamatergic transmission and social behaviour in the mdx52 mouse model of Duchenne muscular dystrophy. Prog Neurobiol 216:102288. https://doi.org/10.1016/j.pneurobio.2022.102288
12. Dixon ML, Thiruchselvam R, Todd R, Christoff K (2017) Emotion and the prefrontal cortex: an integrative review. Psychol Bull 143(10):1033–1081. https://doi.org/10.1037/bul0000096
13. Ko J (2017) Neuroanatomical substrates of rodent social behavior: the medial prefrontal cortex and its projection patterns. Front Neural Circuits 11:41. https://doi.org/10.3389/fncir.2017.00041
14. Gangopadhyay P, Chawla M, Dal Monte O, Chang SWC (2021) Prefrontal-amygdala circuits in social decision-making. Nat Neurosci 24(1):5–18. https://doi.org/10.1038/s41593-020-00738-9
15. Huang WC, Chen Y, Page DT (2016) Hyperconnectivity of prefrontal cortex to amygdala projections in a mouse model of macrocephaly/autism syndrome. Nat Commun 7:13421. https://doi.org/10.1038/ncomms13421
16. Kuniishi H, Nakatake Y, Sekiguchi M, Yamada M (2022) Adolescent social isolation induces distinct changes in the medial and lateral OFC-BLA synapse and social and emotional alterations in adult mice. Neuropsychopharmacology 47(9):1597–1607. https://doi.org/10.1038/s41386-022-01358-6
17. Liu WZ, Zhang WH, Zheng ZH, Zou JX, Liu XX, Huang SH et al (2020) Identification of a prefrontal cortex-to-amygdala pathway for chronic stress-induced anxiety. Nat Commun 11(1):2221. https://doi.org/10.1038/s41467-020-15920-7
18. Kawaminami A, Yamada D, Yoshioka T, Hatakeyama A, Nishida M, Kajino K et al (2024) The delta opioid receptor agonist KNT-127 relieves innate anxiety-like behavior in mice by suppressing transmission from the prelimbic cortex to basolateral amygdala. Neuropsychopharmacol Rep 44(1):256–261. https://doi.org/10.1002/npr2.12406
19. Arruda-Carvalho M, Clem RL (2014) Pathway-selective adjustment of prefrontal-amygdala transmission during fear encoding. J Neurosci 34(47):15601–15609. https://doi.org/10.1523/jneurosci.2664-14.2014
20. Li Y, Missig G, Finger BC, Landino SM, Alexander AJ, Mokler EL et al (2018) Maternal and early postnatal immune activation produce dissociable effects on neurotransmission in mPFC-amygdala circuits. J Neurosci 38(13):3358–3372. https://doi.org/10.1523/jneurosci.3642-17.2018
21. Park K, Chung C (2020) Differential alterations in cortico-amygdala circuitry in mice with impaired fear extinction. Mol Neurobiol 57(2):710–721. https://doi.org/10.1007/s12035-019-01741-3
22. Kauer JA, Malenka RC (2007) Synaptic plasticity and addiction. Nat Rev Neurosci 8(11):844–858. https://doi.org/10.1038/nrn2234

23. Counotte DS, Schiefer C, Shaham Y, O'Donnell P (2014) Time-dependent decreases in nucleus accumbens AMPA/NMDA ratio and incubation of sucrose craving in adolescent and adult rats. Psychopharmacology 231(8): 1675–1684. https://doi.org/10.1007/s00213-013-3294-3

24. Yamada D, Wada K, Sekiguchi M (2016) Modulation of long-term potentiation of cortico-amygdala synaptic responses and auditory fear memory by dietary polyunsaturated fatty acid. Front Behav Neurosci 10:164. https://doi.org/10.3389/fnbeh.2016.00164

Chapter 16

Digital Outcome Measures for Gait and Spontaneous Locomotor Activity in Dystrophic Dogs

Mutsuki Kuraoka, Karin Watanabe, Takahiro Kawashima, Yuko Nitahara-Kasahara, Hisateru Tachimori, Shin'ichi Takeda, Katsura Minegishi, and Yoshitsugu Aoki

Abstract

Digital outcome measures utilizing digital devices and instruments are highly effective in capturing significant changes quantitatively in clinical trials of patients with Duchenne muscular dystrophy. A dystrophic model of beagle dogs, known as canine X-linked muscular dystrophy in Japan ($CXMD_J$), shows the symptoms of muscle weakness and has been evaluated for motor function using these digital outcome measures to assess therapeutic efficacy and monitor disease progression circumstantially. Here, we present accelerometry methods for analyzing gait function and infrared monitoring analysis for assessing spontaneous locomotor activity in $CXMD_J$ dystrophic dogs. The multiple parameters derived from these methods can provide valuable phenotypic information in medium-to-large-sized experimental animals.

Key words Digital outcome measure, Accelerometry, Infrared sensor, Motor function, Gait, Spontaneous locomotor activity, Duchenne muscular dystrophy, Canine X-linked muscular dystrophy

1 Introduction

Duchenne muscular dystrophy (DMD) is an X-linked muscle disorder characterized by primary muscle degeneration [1]. A mutation in the DMD gene results in the absence of dystrophin, a structural protein in muscle fibers, leading to fragile muscle fibers following contractile force [2]. The symptoms of DMD manifest as progressive muscle weakness and contractures, usually appearing as gait disturbance by 2–5 years of age, with the loss of ambulation occurring between 9 and 12 years of age [3, 4]. Steroid therapy for muscle weakness is currently applied in the clinical setting to prolong muscle strength and ambulation; however, steroid therapy is not a cure for DMD [3–5]. Multiple new therapies for DMD, such as gene and cell therapy and pharmaceutical agents, have been explored in clinical trials [6–12].

Yoshitsugu Aoki and Norio Motohashi (eds.), *Splice Switching: Methods and Protocols*, Methods in Molecular Biology, vol. 2963, https://doi.org/10.1007/978-1-0716-4738-7_16,

Outcome measures are needed to evaluate muscle strength and motor function in patients with DMD [6]. Clinical trials must have a defined primary endpoint wherein the study drug results in a significant change in outcomes between treated and untreated patients. Traditional outcome measures, such as timed assessments (e.g., 6-min walking test, running 9 m and rising from the floor) [13–15] and clinical scores (e.g., North Star Ambulatory Assessment, Motor Function Measure and Performance of Upper Limb) [16–18], have been extensively applied to clinical trials but are often less sensitive to disease change (e.g., younger boys with DMD) [19, 20]. Digital outcome measures, which provide mathematical insights utilizing digital devices and instruments, have been reported to have feasibility, reliability, sensitivity to significant changes, and external validity [20–23]. Since the development of miniature wearable sensors, accelerometry has been a highly effective method to quantify performance in validated tasks and activities of daily living [20, 24–26]. Physical activities involving both ambulatory and nonambulatory conditions have been assessed in patients with DMD by measuring acceleration parameters that indicate the movement and orientation of the body and upper limbs [27–31]. Accelerometry for informal tasks such as walking is used to identify deterioration or significant improvement during disease progression or treatment [20, 32–34]. Accelerometry is also more practical for capturing motion when combined with several types of information, such as angular velocity [26, 29, 32].

To explore the efficacy of different nonclinical treatments for DMD, animal models that show dystrophic phenotypes are needed. Canine X-linked muscular dystrophy, which is comparable to severe degenerative myopathy, has been documented in golden retriever muscular dystrophy (GRMD) and is recognized as a genetic homologue of DMD owing to the lack of dystrophin protein [35, 36]. A dystrophic model of beagle dogs, canine X-linked muscular dystrophy in Japan ($CXMD_J$), has been established as a breeding colony by inseminating beagles with the sperm of GRMD dystrophic dogs [37, 38]. These dystrophic dog models show the progressive symptoms of muscle weakness and have been investigated for the quantification of their motor function using digital outcome measures to capture dystrophic phenotypes as well as the efficacy of therapeutic modalities [39–42]. In GRMD, digital outcome measures, including accelerometry and video processing, have revealed reductions in dynamics and kinematic parameters such as speed, stride length, stride frequency, stride regularity, total acceleration power, and joint angles during gait compared with wild-type (WT) dogs [43–48]. Gait in $CXMD_J$ has been evaluated using accelerometry based on multiple parameters (e.g., acceleration and angular velocity), and differences have been reported between sensor-fixed dorsal thoracic and lumbar regions [49]. Activities of daily living have also been assessed in dystrophic dogs by monitoring their spontaneous locomotor activity in a test cage using an infrared monitoring system [44, 49–51]. In this chapter, we present accelerometry methods for analyzing gait function and infrared monitoring analysis for assessing spontaneous locomotor activity in $CXMD_J$ dystrophic dogs.

2 Materials

2.1 Accelerometry for Gait Function

All devices, software, and relevant instructions detailed in this chapter were provided and licensed by ATR-Promotions, Inc. (Soraku-gun, Kyoto, Japan).

1. Portable wireless hybrid sensors: TSND121, TSND151, or AMWS020 (Fig. 1a) (*see* **Note 1**). These sensors have features to detect three-axial acceleration and angular velocity during movement (*see* **Note 2**).

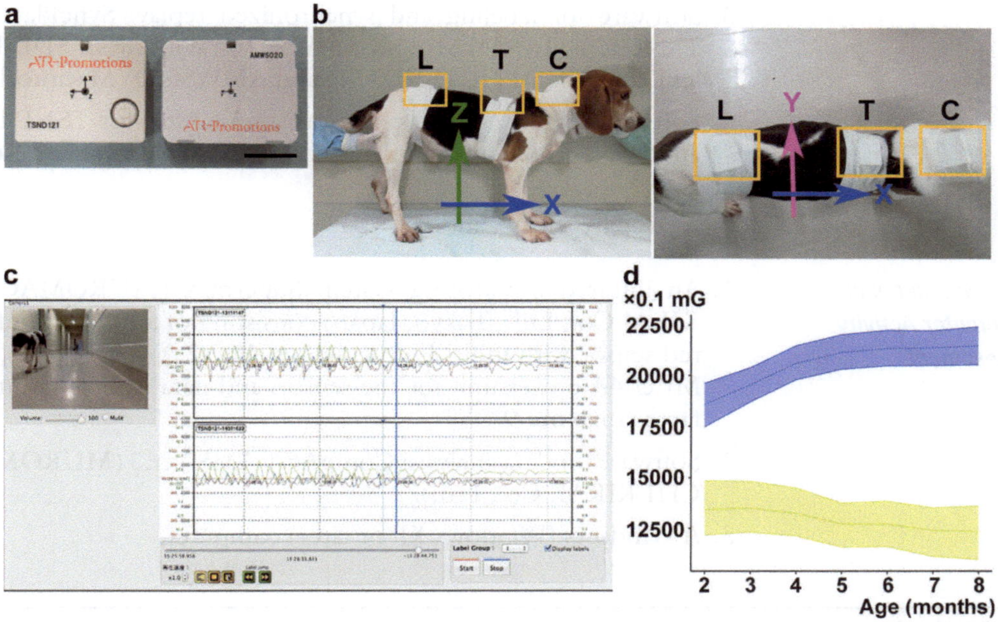

Fig. 1 Accelerometry for gait function in dystrophic dogs. (**a**) Portable sensors TSND121 (left) and AMWS020 (right) for measurement of three-axial acceleration and angular velocity. Scale bar, 20 mm. (**b**) Sensors are worn on the dog's dorsal cervical, thoracic, and lumbar regions. The three axes are the X-axis for the caudal–cranial direction, the Y-axis for the medial–lateral direction, and the Z-axis for the ventral–dorsal direction. C, cervical; T, thoracic; L, lumbar. (**c**) Replaying acceleratory waveforms synchronized with a video showing the gait of a dystrophic dog using SDLabelerT software. (**d**) Average acceleration magnitudes in the lumbar region of wild-type (WT, $n = 9$) and dystrophic dogs ($n = 10$) by months of age. Lines and shaded surrounding areas in WT (blue) and dystrophic dogs (yellow) indicate estimations and 95% credible intervals of the trends

2. A Windows (Windows 7 or later) or Macintosh (OS X 10.6 or later) computer with a built-in or external USB camera, microphone, and Bluetooth.
3. Software for data reception synchronized with video and sound, recording, viewing, and replay: ALTIMA for TSND121, TSND151, and AMWS020 (Windows). SDRecorderT for TSND121 and TSND151 (Macintosh). Software for data reception, recording, and graphical display: SensorController for TSND121, TSND151, and AMWS020 (Windows).
4. Charge-free download tools (https://www.atr-p.com/support/TSND-tools.html): USB Serial Driver for TSND121, TSND151, and AMWS020 (Windows) (*see* **Note 3**). Bluetooth pairing tool for TSND121, TSND151, and AMWS020 (Windows) (*see* **Note 4**). DataDownloader for TSND121, TSND151, and AMWS020 (Windows) or TSND121 and TSND151 (Macintosh) (*see* **Note 5**). DataConverter for TSND121, TSND151, and AMWS020 (Windows) (*see* **Note 6**).
5. Software for labeling and synchronized replay: SyncPlay for TSND121, TSND151, and AMWS020 (Windows). SDLabelerT for TSND121, TSND151, and AMWS020 (Macintosh).
6. Software for data analysis: SensorDataAnalyzer for TSND121 and TSND151, and partially AMWS020 (Windows) (*see* **Note 7**).
7. A stretchy sticky bandage.

2.2 Spontaneous Locomotor Activity Assessment

1. An infrared monitoring system Supermex (MUROMACHI KIKAI Co., Ltd., Tokyo, Japan) consisting of a PYS-001 infrared sensor with multi-Fresnel lenses (Fig. 2a) (*see* **Note 8**), an MPU-4A power unit for the sensor, and data recording interfaces (*see* **Note 9**).
2. Software for data analysis: CompACT AMS Ver.3 (MUROMACHI KIKAI, Co., Ltd.).
3. A Windows (Windows XP or later) computer.

3 Methods

3.1 Accelerometry for Gait Function

1. Affix portable sensors on the dorsal position of subject dogs with a stretchy sticky bandage (*see* **Notes 10** and **11**). For instance, sensors are worn around the epistropheus, sixth thoracic, and seventh lumbar vertebrae of subjects to assess the movement of the dorsal cervical, thoracic, and lumbar regions, respectively (Fig. 1b).

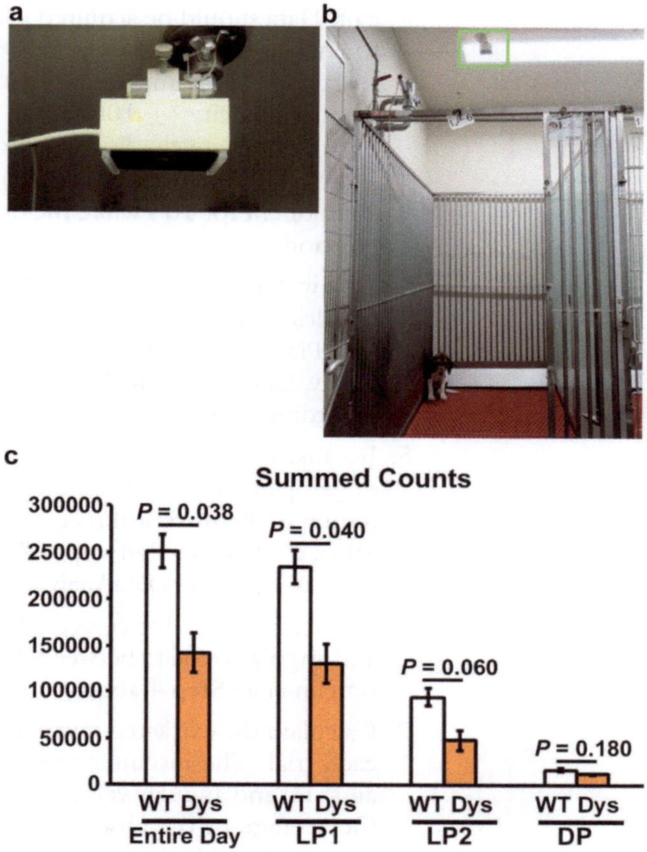

Fig. 2 Infrared monitoring for spontaneous locomotor activity. (**a**) A PYS-001 infrared sensor with multi-Fresnel lenses. (**b**) Setting up to mount above a dog test cage. The PYS-001 infrared sensor is shown in the green box. (**c**) Comparison of summed counts of spontaneous locomotor activity between wild-type (WT, $n = 3$) and dystrophic (Dys, $n = 3$) dogs, according to different periods. Light period (LP), time between 07:00 and 19:00. Dark period (DP), time between 19:00 and 07:00 the next day. LP1, the entire light period. LP2, the light period excluded the broad time window during which animal keepers were feeding the dogs or cleaning cages. Data are expressed as mean ± SEM. *P-values were compared between WT and dystrophic dogs using Welch's t-test. All calculations were performed using R version 4.1.0 (https:/www.r-project.org)

2. For data recording, select the online (*see* **Note 4**) or offline setting. The online setting enables data recording using SensorController or ALTIMA (Windows), or SDRecorderT (Macintosh). In the offline setting, data are recorded in the built-in memory of sensors and downloaded later (*see* **Note 5**).
3. When performing the 15 m running test, in which a dog runs down a hallway (15 m × 4 times) (*see* **Note 12**) [49, 50, 52], measure three-axial acceleration and angular velocity during

gait. Data should be acquired using ALTIMA or SDRecorderT to enable synchronization with sound and video, viewing and replaying. We sampled three-axial acceleration (± 8G) and angular velocity (± 1000 dps) at 0.24mG and 0.03 dps per 20 ms, respectively. For data correction by removal of offset (gravity), as mentioned in step "5", record a direct current component for 10 s while the dog is in a quadrupedal standing position.

4. As raw data recorded using ALTIMA and SDRecorderT, wave profiles synchronized with sound and video can be replayed in SyncPlay and SDLabelerT, respectively (Fig. 1c). During replay, label the "start" and "goal" times in each running trial according to a timeline.
5. Because raw acceleration data include gravity regardless of the dogs' gait, data correction by offset (gravity) removal is required. Remove a direct current component (recorded for 10 s), as mentioned in step "3″ above, from all raw acceleratory data using SensorDataAnalyzer (*see* **Notes 13** and **14**).
6. Extract raw data for corrected acceleration (*see* **Step** 5 above) and angular velocity between "start" and "goal" on the labeled timeline (*see* **Step** 4 above).
7. Calculate the extracted raw data using the operations below, in each trial. The instantaneous vectors of acceleration (*Ax*, *Ay*, and *Az*) and angular velocity (*Gx*, *Gy*, and *Gz*) are calculated as the average of the absolute values for each axis. Acceleration magnitude (*AM*) is calculated from the three acceleration vectors (*Ax*, *Ay*, and *Az*) as the square root of the sum of the three-axial values ($AM = \sqrt{Ax^2 + Ay^2 + Az^2}$) [53] and then averaged for each trial. The relative components of *AM* along the three axes (%) are calculated by dividing the absolute values for each axis by *AM* [46], and the components that are averaged in each trial are calculated as acceleration ratios (*Ax* ratio, *Ay* ratio, and *Az* ratio).
8. When conducting longitudinal measurements of acceleration and angular velocity, time-series analysis is applicable for comparisons between dystrophic and WT dogs (Fig. 1d). For instance, we estimated trends in the multiple parameters of acceleration and angular velocity as the following local-level models. In these models, i and t are indices of "dog" and "month", respectively. For acceleration or angular velocity, we assumed the following model: $\log(y_{it}) \sim N(m_t, \sigma_0^2)$. For the acceleration ratio, we assumed the following model: $\log\left(\frac{r_{it}}{100 - r_{it}}\right) \sim N(m_t, \sigma_0^2)$, where y_{it} is acceleration or angular velocity, r_{it} is the acceleration ratio, and m_t is the trend. For m_t in the above two models, we assumed the following model: $m_t \sim N(m_{t-1}, {\sigma_1}^2)$.

3.2 Spontaneous Locomotor Activity Assessment

1. An infrared sensor is mounted above a dog test cage (Fig. 2b) to detect animal locomotion as changes in heat across multiple zones of the cage using an array of Fresnel lenses [50, 51]. In this way, the system automatically monitors and counts all spontaneous movements, both vertical and horizontal, of the dog.
2. Keep all dog subjects under isolation and reared in a test cage (*see* **Note 15**).
3. Set up the monitoring conditions, including the sampling intervals (*see* **Note 16**), monitoring terms, and number of channels to monitor plural subjects in parallel. For instance, we sampled data every 1–10 min over 5 days.
4. Record all count data using CompACT AMS software Ver. 3 and export the results in CSV file format.
5. Apply raw count data to a comparison analysis between dystrophic and WT dogs and calculate the sum or average values in the term. Extract appropriate raw count data in a timeline according to the research applications (e.g., light and dark periods) (Fig. 2c).

4 Notes

1. TSND121 is not for sale at the present time. Alternatively, TSND151 and AMWS020 are treated as succession devices. The specifications of each sensor are as follows: TSND121, size: 46 mm (W) × 37 mm (H) × 12 mm (D), weight: 22 g, acceleration range ± 2, ±4, ±8, and ± 16G, angular velocity range: ±250, ±500, ±1000, and ± 2000 dps, data sampling per 1–255 ms and up to 1000 Hz, operation-time: 6 h; TSND151, size: 40 mm (W) × 50 mm (H) × 14 mm (D), weight: 27 g, acceleration range: ±2, ±4, ±8, and ± 16G, angular velocity range: ±250, ±500, ±1000, and ± 2000 dps, data sampling per 1–255 ms and up to 1000 Hz, operation-time: 8.7 h; AMWS020, size: 37 mm (W) × 46 mm (H) × 12 mm (D), weight: 22 g, acceleration range: ±4, ±8, ±16, and ± 30G (enabled to be equipped with additional ±200G), angular velocity range: ±500, ±1000, ±2000, and ± 4000 dps, data sampling per 1–255 ms and up to 1000 Hz or per 0.25 ms and up to 4000 Hz, operation time: 6 h. When detecting acceleration, resolution powers (RPs) of acceleration 0.06, 0.12, 0.24, 0.48, and 0.98 mG correspond to acceleration ranges ±2, ±4, ±8, ±16, and ± 30G, respectively. When detecting angular velocity, RPs of angular velocity 0.008, 0.015, 0.030, 0.061, and 0.122 dps correspond to

angular velocity ranges ±250, ±500, ±1000, ±2000, and ± 4000 dps, respectively.

2. The specific acceleration vector indicates the instantaneous inertial acceleration for each axis (*Ax*, *Ay*, and *Az*) and is expressed in G-force (1G = 9.81 m/s^2). The instantaneous angular velocity vector indicates the instantaneous rotation of the axes (*Gx*, *Gy*, and *Gz*) and is expressed in dps.
3. Before connecting sensors to a computer through a wired connection, a USB serial driver must be installed.
4. A Bluetooth pairing tool is convenient for connecting sensors using a Windows computer with a Bluetooth connection, although it is not necessary. The computer must be equipped with Bluetooth Ver2.0 + EDR Class1 or 2.
5. When downloading data from built-in sensor memory, use DataDownloader, ALTIMA, or SensorController (Windows) or DataDownloader (Macintosh).
6. DataConverter is used to convert the clock format and angle of inclination and classify the data, including “acceleratory and angular velocity,” “geomagnetism,” and “atmosphere and temperature,” in CSV file format, following export from SensorController, ALTIMA, and SDRecorderT. When using SyncPlay data-labeling software, DataConverter is needed to convert the CSV file to the dedicated format.
7. At present, SensorDataAnalyzer can be partially applied to AMWS020 (limited to the analysis of compatible data with TSND121 and TSND151).
8. The infrared sensor is sized 55 mm (W) × 75 mm (H) × 85 mm (D) and weighs 100 g.
9. DI-008 W, DI-016 W, DI-032 W, and DI-064 W interfaces are capable of recording data on 8, 16, 32, and 64 channels, respectively.
10. Insufficient fixation causes separation between the sensor and subject body during gait, leading to interference and a lot of noise. It is important to fix the sensors thoroughly using a sticky bandage, while taking care not to tighten and distress the animals excessively. It is recommended that the subjects be adequately adjusted to a bundled state prior to a trial.
11. The directions of the three axes of acceleration and angular velocity must be determined in advance, and the directions should be made consistent when using multiple sensors and/or conducting longitudinal measurements. For instance, the three axes are set as the X-axis (caudal–cranial), Y-axis (medial–lateral), and Z-axis (ventral–dorsal) (Fig. 1b) [49]. The sensor locations are selected according to the base points of motion and/or center of mass.

12. It is recommended that dogs be adjusted in advance to a running task and a trial environment. Dogs tend to run naturally when accompanied by a familiar animal keeper. Consider stopping the trial based on the condition of the dog (e.g., sitting down as a result of fatigue).
13. A direct current component can be set as an initial offset and removed by a low-cut filter using SensorDataAnalyzer software.
14. SensorDataAnalyzer software is also applicable to physics calculations (e.g., velocity, angle, and locomotion distance) and frequency analysis (e.g., fast Fourier transformation).
15. All subject dogs are adjusted in advance to an environment under isolation rearing in a test cage.
16. Data acquisition can be set per 1–60 min.

Conflict of Interest

The authors have no conflicts of interest to declare that are relevant to the content of this chapter.

References

1. Sussman M (2002) Duchenne muscular dystrophy. J Am Acad Orthop Surg 10(2): 138–151
2. Matsumura K, Campbell KP (1994) Dystrophin-glycoprotein complex: its role in the molecular pathogenesis of muscular dystrophies. Muscle Nerve 17(1):2–15. https://doi.org/10.1002/mus.880170103
3. Goto M et al (2016) Long-term outcomes of steroid therapy for Duchenne muscular dystrophy in Japan. Brain Dev 38(9):785–791. https://doi.org/10.1016/j.braindev.2016.04.001
4. Manzur AY, Kuntzer T, Pike M, Swan A (2008) Glucocorticoid corticosteroids for Duchenne muscular dystrophy. Cochrane Database Syst Rev 1:CD003725. https://doi.org/10.1002/14651858.CD003725.pub3
5. Guiraud S et al (2015) The pathogenesis and therapy of muscular dystrophies. Annu Rev Genomics Hum Genet 16:281–308. https://doi.org/10.1146/annurev-genom-090314-025003
6. Shieh PB (2015) Duchenne muscular dystrophy: clinical trials and emerging tribulations. Curr Opin Neurol 28(5):542–546. https://doi.org/10.1097/WCO.0000000000000243
7. Dang UJ et al (2024) Efficacy and safety of Vamorolone over 48 weeks in boys with Duchenne muscular dystrophy: a randomized controlled trial. Neurology 102(5):e208112. https://doi.org/10.1212/WNL.0000000000208112
8. Mercuri E et al (2024) Safety and efficacy of givinostat in boys with Duchenne muscular dystrophy (EPIDYS): a multicentre, randomised, double-blind, placebo-controlled, phase 3 trial. Lancet Neurol 23(4):393–403. https://doi.org/10.1016/S1474-4422(24)00036-X
9. Mendell JR et al (2025) AAV gene therapy for Duchenne muscular dystrophy: the EMBARK phase 3 randomized trial. Nat Med 31(1): 332–341. https://doi.org/10.1038/s41591-024-03304-z
10. Henzi BC et al (2024) Tamoxifen may contribute to preserve cardiac function in Duchenne muscular dystrophy. Eur J Pediatr 183(9): 4057–4062. https://doi.org/10.1007/s00431-024-05670-9
11. Lee J et al (2025) Safety and tolerability of Wharton's jelly-derived mesenchymal stem cells for patients with Duchenne muscular dystrophy: a phase 1 clinical study. J Clin Neurol

21(1):40–52. https://doi.org/10.3988/jcn.2024.0299

12. Komaki H et al (2025) Phase 1/2 trial of brogidirsen: dual-targeting antisense oligonucleotides for exon 44 skipping in Duchenne muscular dystrophy. Cell Rep Med 6(1):101901. https://doi.org/10.1016/j.xcrm.2024.101901
13. McDonald CM et al (2013) The 6-minute walk test and other endpoints in Duchenne muscular dystrophy: longitudinal natural history observations over 48 weeks from a multicenter study. Muscle Nerve 48(3):343–356. https://doi.org/10.1002/mus.23902
14. McDonald CM et al (2013) The 6-minute walk test and other clinical endpoints in duchenne muscular dystrophy: reliability, concurrent validity, and minimal clinically important differences from a multicenter study. Muscle Nerve 48(3):357–368. https://doi.org/10.1002/mus.23905
15. Beenakker EA et al (2005) Functional ability and muscle force in healthy children and ambulant Duchenne muscular dystrophy patients. Eur J Paediatr Neurol 9(6):387–393. https://doi.org/10.1016/j.ejpn.2005.06.004
16. Mazzone ES et al (2009) Reliability of the north star ambulatory assessment in a multicentric setting. Neuromuscul Disord 19(7):458–461. https://doi.org/10.1016/j.nmd.2009.06.368
17. Schreiber A et al (2018) Corticosteroids in Duchenne muscular dystrophy: impact on the motor function measure sensitivity to change and implications for clinical trials. Dev Med Child Neurol 60(2):185–191. https://doi.org/10.1111/dmcn.13590
18. Pane M et al (2014) Reliability of the performance of upper limb assessment in Duchenne muscular dystrophy. Neuromuscul Disord 24(3):201–206. https://doi.org/10.1016/j.nmd.2013.11.014
19. Lu QL, Cirak S, Partridge T (2014) What can we learn from clinical trials of exon skipping for DMD? Mol Ther Nucleic Acids 3:e152. https://doi.org/10.1038/mtna.2014.6
20. Gonzalez Barral C, Servais L (2024) Digital outcome measures in Duchene muscular dystrophy: lessons learnt from clinical trials. J Neuromuscul Dis:22143602241296280. https://doi.org/10.1177/22143602241296280
21. Ferrer-Mallol E et al (2022) Patient-led development of digital endpoints and the use of computer vision analysis in assessment of motor function in rare diseases. Front Pharmacol 13:916714. https://doi.org/10.3389/fphar.2022.916714
22. Rabbia M et al (2024) Stride velocity 95th centile detects decline in ambulatory function over shorter intervals than the 6-minute walk test or north star ambulatory assessment in Duchenne muscular dystrophy. J Neuromuscul Dis 11(3):701–714. https://doi.org/10.3233/JND-230188
23. Servais L et al (2023) First regulatory qualification of a digital primary endpoint to measure treatment efficacy in DMD. Nat Med 29(10):2391–2392. https://doi.org/10.1038/s41591-023-02459-5
24. Neugebauer JM, Hawkins DA, Beckett L (2012) Estimating youth locomotion ground reaction forces using an accelerometer-based activity monitor. PLoS One 7(10):e48182. https://doi.org/10.1371/journal.pone.0048182
25. Rowlands AV et al (2015) Comparability of measured acceleration from accelerometry-based activity monitors. Med Sci Sports Exerc 47(1):201–210. https://doi.org/10.1249/MSS.0000000000000394
26. Salarian A et al (2007) Ambulatory monitoring of physical activities in patients with Parkinson's disease. IEEE Trans Biomed Eng 54(12):2296–2299
27. Ganea R et al (2012) Gait assessment in children with duchenne muscular dystrophy during long-distance walking. J Child Neurol 27(1):30–38. https://doi.org/10.1177/0883073811413581
28. Davidson ZE et al (2015) Strong correlation between the 6-minute walk test and accelerometry functional outcomes in boys with Duchenne muscular dystrophy. J Child Neurol 30(3):357–363. https://doi.org/10.1177/0883073814530502
29. Le Moing AG et al (2016) A movement monitor based on magneto-inertial sensors for non-ambulant patients with Duchenne muscular dystrophy: a pilot study in controlled environment. PLoS One 11(6):e0156696. https://doi.org/10.1371/journal.pone.0156696
30. Kimura S et al (2014) Estimation of muscle strength from actigraph data in Duchenne muscular dystrophy. Pediatr Int 56(5):748–752. https://doi.org/10.1111/ped.12348
31. Ramli AA et al (2024) Gait characterization in Duchenne muscular dystrophy (DMD) using a single-sensor accelerometer: classical machine learning and deep learning approaches. Sensors (Basel) 24(4). https://doi.org/10.3390/s24041123

32. Jeannet PY et al (2011) Continuous monitoring and quantification of multiple parameters of daily physical activity in ambulatory Duchenne muscular dystrophy patients. Eur J Paediatr Neurol 15(1):40–47. https://doi.org/10.1016/j.ejpn.2010.07.002
33. Nishizawa H, Shiba N, Nakamura A (2016) Usefulness of continuous actigraph monitoring in the assessment of the effect of corticosteroid treatment for Duchenne muscular dystrophy: a case report. J Phys Ther Sci 28(11):3249–3251. https://doi.org/10.1589/jpts.28.3249
34. Killian M et al (2020) Beyond ambulation: measuring physical activity in youth with Duchenne muscular dystrophy. Neuromuscul Disord 30(4):277–282. https://doi.org/10.1016/j.nmd.2020.02.007
35. Kornegay JN et al (2012) Canine models of Duchenne muscular dystrophy and their use in therapeutic strategies. Mamm Genome 23(1–2):85–108. https://doi.org/10.1007/s00335-011-9382-y
36. Kornegay JN (2017) The golden retriever model of Duchenne muscular dystrophy. Skelet Muscle 7(1):9. https://doi.org/10.1186/s13395-017-0124-z
37. Shimatsu Y et al (2003) Canine X-linked muscular dystrophy in Japan (CXMDJ). Exp Anim 52(2):93–97
38. Kuraoka M, Aoki Y, Takeda S (2021) Development of outcome measures according to dystrophic phenotypes in canine X-linked muscular dystrophy in Japan. Exp Anim 70(4):419–430. https://doi.org/10.1538/expanim.21-0072
39. Valentine BA, Cooper BJ, Cummings JF, deLahunta A (1986) Progressive muscular dystrophy in a golden retriever dog: light microscope and ultrastructural features at 4 and 8 months. Acta Neuropathol 71(3–4):301–310. https://doi.org/10.1007/BF00688053
40. Valentine BA et al (1988) Canine X-linked muscular dystrophy. An animal model of Duchenne muscular dystrophy: clinical studies. J Neurol Sci 88(1–3):69–81. https://doi.org/10.1016/0022-510x(88)90206-7
41. Shimatsu Y et al (2005) Major clinical and histopathological characteristics of canine X-linked muscular dystrophy in Japan. CXMDJ Acta Myol 24(2):145–154
42. Kornegay JN et al (2003) The cranial sartorius muscle undergoes true hypertrophy in dogs with golden retriever muscular dystrophy. Neuromuscul Disord 13(6):493–500. https://doi.org/10.1016/s0960-8966(03)00025-7
43. Marsh AP et al (2010) Kinematics of gait in golden retriever muscular dystrophy. Neuromuscul Disord 20(1):16–20. https://doi.org/10.1016/j.nmd.2009.10.007
44. Shin JH et al (2013) Quantitative phenotyping of Duchenne muscular dystrophy dogs by comprehensive gait analysis and overnight activity monitoring. PLoS One 8(3):e59875. https://doi.org/10.1371/journal.pone.0059875
45. Barthelemy I et al (2009) Gait analysis using accelerometry in dystrophin-deficient dogs. Neuromuscul Disord 19(11):788–796. https://doi.org/10.1016/j.nmd.2009.07.014
46. Barthelemy I et al (2011) Longitudinal ambulatory measurements of gait abnormality in dystrophin-deficient dogs. BMC Musculoskelet Disord 12:75. https://doi.org/10.1186/1471-2474-12-75
47. Barthelemy I et al (2014) Predictive markers of clinical outcome in the GRMD dog model of Duchenne muscular dystrophy. Dis Model Mech 7(11):1253–1261. https://doi.org/10.1242/dmm.016014
48. Fraysse B et al (2017) Gait characterization in golden retriever muscular dystrophy dogs using linear discriminant analysis. BMC Musculoskelet Disord 18(1):153. https://doi.org/10.1186/s12891-017-1494-4
49. Kuraoka M et al (2018) Accelerometric outcomes of motor function related to clinical evaluations and muscle involvement in dystrophic dogs. PLoS One 13(12):e0208415. https://doi.org/10.1371/journal.pone.0208415
50. Hayashita-Kinoh H et al (2015) Intra-amniotic rAAV-mediated microdystrophin gene transfer improves canine X-linked muscular dystrophy and may induce immune tolerance. Mol Ther 23(4):627–637. https://doi.org/10.1038/mt.2015.5
51. Echigoya Y et al (2017) Effects of systemic multiexon skipping with peptide-conjugated morpholinos in the heart of a dog model of Duchenne muscular dystrophy. Proc Natl Acad Sci USA 114(16):4213–4218. https://doi.org/10.1073/pnas.1613203114
52. Yokota T et al (2009) Efficacy of systemic morpholino exon-skipping in Duchenne dystrophy dogs. Ann Neurol 65(6):667–676. https://doi.org/10.1002/ana.21627
53. Galan-Mercant A, Baron-Lopez FJ, Labajos-Manzanares MT, Cuesta-Vargas AI (2014) Reliability and criterion-related validity with a smartphone used in timed-up-and-go test. Biomed Eng Online 13:156. https://doi.org/10.1186/1475-925X-13-156

INDEX

Yoshitsugu Aoki and Norio Motohashi (eds.), *Splice Switching: Methods and Protocols*, Methods in Molecular Biology, vol. 2963, https://doi.org/10.1007/978-1-0716-4738-7,

MIX
Papier aus verantwortungsvollen Quellen
Paper from responsible sources
FSC® C105338

If you have any concerns about our products,
you can contact us on
ProductSafety@springernature.com

In case Publisher is established outside the EU,
the EU authorized representative is:
Springer Nature Customer Service Center GmbH
Europaplatz 3, 69115 Heidelberg, Germany

Printed by Libri Plureos GmbH
in Hamburg, Germany